THE BEGINNING OF HEAVEN
AND EARTH HAS NO NAME

MEANING SYSTEMS

THE BEGINNING OF HEAVEN AND EARTH HAS NO NAME

Seven Days with Second-Order Cybernetics

HEINZ VON FOERSTER

EDITED BY ALBERT MÜLLER
AND KARL H. MÜLLER

TRANSLATED BY ELINOR ROOKS
AND MICHAEL KASENBACHER

Fordham University Press : New York 2014

Frontispiece: Heinz von Foerster at his home, Pescadero, California, in 1997

The Beginning of Heaven and Earth Has No Name was published in German as *Der Anfang von Himmel und Erde hat keinen Namen. Eine Selbsterschaffung in 7 Tagen*, © Kulturverlag Kadmos Berlin, 2002.

Library of Congress Cataloging-in-Publication Data

Von Foerster, Heinz, 1911–2002, author.
    [Anfang von Himmel und Erde hat keinen Namen. English]
    The beginning of heaven and Earth has no name : seven days with second-order cybernetics / Heinz von Foerster ; edited by Albert Müller and Karl H. Müller ; translated by Elinor Rooks and Michael Kasenbacher. — First edition.
        pages cm. — (Meaning systems)
    Includes bibliographical references and index.
    ISBN 978-0-8232-5560-3 (hardback) —
    ISBN 978-0-8232-5561-0 (paper)
    1. Cybernetics. 2. Knowledge, Theory of. 3. Science—Philosophy. I. Müller, Albert, 1959– editor. II. Müller, Karl H., editor. III. Title.
    Q310.V6613 2014
    003'.5—dc23

                          2013030580

16 15 14 5 4 3 2 1

First edition

CONTENTS

A FOREWORD BY THE SERIES EDITOR

Heinz von Foerster is one of the most consequential cybernetic thinkers in the history of the field. He was born in 1911 in Vienna, Austria, into a progressive bourgeois family of architects, designers, artists, and activists. Hailing from a partially Jewish background, he weathered the Nazi era by moving with his wife from recognition in Vienna to obscurity in Berlin. His university studies in physics enabled him to secure employment in corporate research laboratories. After repatriation at the end of World War II, Foerster worked at a telephone company and as a commentator for a radio station operated by the American military. In 1949, he came to the United States seeking employment. He gained the attention of Warren McCulloch, who subsequently brought him into the later meetings of the Macy Conferences on Cybernetics and assisted his securing an academic position in the Electrical Engineering department at the University of Illinois. In 1957 he established his own research program there, the Biological Computer Laboratory, and directed it until it closed and he retired in 1975. A series of seminal colleagues and collaborators came to the BCL, among them Ross Ashby, Gordon Pask, Gotthard Günther, Lars Löfgren, John Lilly, and Humberto Maturana. After retirement and almost until his death in 2002, Foerster maintained an active round of conference attendance and professional speaking engagements.

In all that time, although he was the author or coauthor of nearly two hundred professional papers, and despite his growing renown and significance within a variety of fields, he never wrote a book. Nor—unlike, for instance, Gregory Bateson, who did so in *Mind and Nature*—did Foerster gather his thinking up into a sustained summary statement. This unique work, *The Beginning of Heaven and Earth Has No Name*, has come into existence to redress the lack in Foerster's own oeuvre of such a book, and in particular, to address the need for an accessible, nonmathematical, comprehensive overview of his cybernetic ideas and the philosophy latent within them. This book gathers into a single volume a certain distillation of concepts and projects scattered amidst a life's work left in short or single-topic statements—mostly professional papers, and most of them published versions of spoken addresses. As you will discover, its editors have endowed it with the coherence and productive linearity of a programmatic framework. Nonetheless, this is still an interview book, and as such it does justice to Foerster's élan as a speaker and improviser. His forte as a raconteur has not been diminished by this treatment.

If one does consult collections of his papers (which others, such as Francisco Varela, had to bring into being), one typically finds a mix of sophisticated wit and complicated mathematics. In this volume, although math is often discussed, it does not intrude its immediate methods into the conversation. Here we have a fully discursive Foerster, which is admittedly an artifact of the editors' dialogical method. In his retirement, the Austrian émigré was once again repatriated by the city of his birth and upbringing. Published almost entirely in English, his work was taken up and championed by a range German-speaking academics and thinkers, including those developing radical constructivism and others following the lead of Niklas Luhmann's social systems theory. Directed by coeditor Albert Müller, a Heinz von Foerster Archive has been established at the University of Vienna, while both coeditors, Albert Müller and Karl H. Müller, have been instrumental in developing the Heinz von Foerster Society and organizing a biannual congress. Originally developed by these Austrian editors for a German press, this American edition of *The Beginning of Heaven and Earth Has No Name* will allow English speakers of all descriptions a new ease of access to the rich thought of this remarkable and protean scientist.

BRUCE CLARKE

AN AUTHOR'S FOREWORDS

Draw a Distinction!
—G. SPENCER BROWN, cited by Heinz von
    Foerster, *On Constructing a Reality*

The trouble about progress is that it always
looks much greater than it really is.
—NESTROY, cited by Ludwig Wittgenstein,
    motto of *Philosophical Investigations*

Freely I have received, freely I have given,
and I request nothing for it.
—MARTIN LUTHER, Prologue to his translation of the Bible

When it came to writing a book, I was corrupted very early on by two doctrines from Ludwig Wittgenstein's *Tractatus Logico-Philosophicus*. These are the first and the last sentences of his *Tractatus*. The first is a quotation from Ferdinand Kürnberger, which Wittgenstein put in as a motto for the whole work: *and whatever a man knows, whatever is not mere rumbling and roaring that he has heard, can be said in three words.* The second doctrine is the famous Proposition 7, the final sentence of his *Tractatus*: *Whereof one cannot speak, thereof one must be silent.*

The effect of these admonishments on me was one of singular self-constraint: whenever I dared to write a sentence, I would count up the words in it, and there would be more than three. Thus I began to shorten and cut, but there were always more than three words. Of course, I could have just as well discounted the Kürnberger sentence, too, because that also had more than three words, but I wanted to move *within* the framework of the *Tractatus* for a while. Eventually I would come to suspect that my sentence was a case for Proposition 7—and thereof I must be silent. Then I would tackle the next sentence. But you can easily imagine the fate of this sentence: it had to go as well. On to the next, more than three words there, too, "Kürnberger's Razor," gone . . . So there were fewer and fewer sentences, or maybe none at all. "And when the emperor falls the duke must follow suit"—and when the sentences disappear, there are no grounds for finishing a *book*.[1]

Much later, I thought of a third principle, which clipped my literary wings still further. I called it the "hermeneutical principle": *The listener, not the speaker, determines the meaning of an utterance.* Then one day, as if conjured by a fairy, two spirits appeared—with one name, of course, "Müller"— and said, "*Heinz, we want to take you seriously!*" What? How do you mean to do that? "*Do you remember your hermeneutical principle about listeners and speakers? In a literary context, it becomes, 'The reader, not the writer, determines the meaning of a sentence,' doesn't it?*" Yes, and? "*You have to give the reader something to interpret. 'One can learn even from the most stupid' is another of your propositions. Why won't you finally apply your principles to yourself?*"

It was clear that I had met my match. I had some new masters.

The following pages, the "Days of Creation," as my masters call them— for me, also the "Days of Exhaustion"—are the records of a fish's desperate attempts to get off the hook. As the patient reader will see, the fish did not succeed. Its hope, which it never lost for all its wriggling, was its faith in its masters' ability to transform the drama of the wriggling fish into bait for the reader, who would take pleasure in the meaning that he or she gives this drama.

HEINZ VON FOERSTER, Pescadero, October 1997

I have written countless prefaces for all sorts of things, but the best preface I've ever written is the one for *The Beginning of Heaven and Earth*.

HEINZ VON FOERSTER, Pescadero, June 2001

FOREWORDS WITH TWO EDITORS

> If we don't act ourselves, we shall be acted upon.
> —HEINZ VON FOERSTER, Understanding Understanding

> When the answer cannot be put into words, neither
> can the question be put into words. . . . If a question
> can be framed at all, it is also possible to answer it.
> —LUDWIG WITTGENSTEIN, Tractatus Logico-Philosophicus

> These are the generations of the heauens, & of the earth,
> when they were created; in the day that the Lord God made
> the earth, and the heauens, And euery plant of the field,
> before it was in the earth, and euery herbe of the field, before
> it grew: for the Lord God had not caused it to raine vpon
> the earth, and there was not a man to till the ground.
> —GENESIS 2:4–6

Finding suitable introductions proves to be quite difficult, if only because incomparably more first sentences present themselves than can
possibly be set down. As a rule, books only have the option of one first
sentence, one single first paragraph. Composing this introduction proved
especially time-consuming because many equally attractive variations
were conceivable. The most immediate and direct form of introduction would refer to the book's intent, aims and contents. This seemed

imperative in our case because there needs to be a careful explanation of what plan, if any, guides this book. The basic idea for the book can be plainly and simply stated: The life's work of Heinz von Foerster, a biocybernetician from the former "greater Austria," should be discussed, along with his specific historical and scientific-historical contexts. The potpourri of Foersterian themes should be arranged and composed, however, in the context of a fundamental problem of biocybernetics—the construction of a "thought machine." In other words, we wanted to apply our "trialogs" with Heinz von Foerster to the question of what steps, devices, heuristics, programs, and hardware would be necessary to construct a "thought machine" that—and here we get onto a strong self-referential loop—would think and function like Heinz von Foerster himself. In this spirit we designed the conversations according to the following schedule:

Monday: Matter
Tuesday: Life
Wednesday: Sense and movement
Thursday: Cognition
Friday: Language
Saturday: Foersterian heuristics
Sunday: Rest

At the happy ending of these six conversations—the lucky number seven has long been associated with rest and contemplation—we should have an overview of the Foersterian perspective on biocybernetics, as well as an insight into the working method and operations of the Foersterian "thought style."

Such an introduction, if carried out in greater detail, would afford insights into the book's design. But because this introduction would force us to disregard so many interesting points, we looked for an alternative framework that would allow us a greater variety and diversity of themes. And almost effortlessly, an introduction with a stronger orientation toward the social sciences emerged, based on the idea of the pundit. Interviews with pundits do not, as a rule, last six long days but make do with a shorter space of time. But pundits have their talking points and programs, which we had also—in repeated consultation with Heinz von Foerster—drafted and coordinated. The emphasis on these talking points will be briefly laid out here and reproduced elsewhere. Throughout the days, each of these six conversations should, like trivial machines, progress through the following internal stages or key passages:

1. Foerster quotation to start the day
2. Key concept of the day
3. Key theorem of the day
4. "Second-order considerations"
5. Key heuristics concerning the field of conversation
6. "Implicit knowledge" of the particular field
7. Foerster quotation to end the day

There should be no pretence that agreement on this program on this conversational program was a foregone conclusion. Indeed, in the interview, Heinz himself accepted the possibility of a distinction between "science" and "life":

There are two main areas that develop within every person who works in science: namely, his personal life, his personal stories, how he came upon his themes, how he decided upon his special subjects, how he met his beautiful wife, when his children were born, and why he had that fierce argument with the great philosopher X, Y, or Z. That's the first aspect, a focus within his personality. The second main area is what he was arguing about with the great philosopher. What concepts were presented? Why was he interested in population growths? Why was he working on haematology? And why the focus on cognition, physiology, and neurology? What did he learn there, what did he see? And what does he claim to know now? Those are the two aspects.

Nevertheless, Heinz—rightly enough—did not want to apply this distinction between science and life to himself: "I would like to keep it so that the connection between the person and science remains central." After some transatlantic back and forth, we had agreed upon a main thread, though it was a high-tension one: "the self-creation of Heinz von Foerster in seven days." Until the end—and probably to this day—the self-creator in question, Heinz von Foerster, remained unconvinced that such an undertaking could be sensibly accomplished. We, on the other hand, were firmly convinced that the seven planned guiding steps would lead to a reasonable approximation of our goal. The possible doubts of our otherwise so productive journeyman will here be afforded only a marginal position.

This sort of introduction soon seemed too fixated on the expert. We tried another scientific possibility—still very cerebral and content-heavy: an introduction to the core themes of this conversational book, namely, to the areas of "cognition and construction." Along this route, we had plenty of points of connection to the relevant cognition theories and constructivist literature. A short review of the relevant precursors in the self-creation of people—not counting alchemy and science fiction—stretches from Valentino Braitenberg's *Vehicles: Experiments in Synthetic Psychology*, an amusing biocybernetic treatise on ever more complicated and intelligent machines, on to John Pollock's *How to Build a Person*, in which the principle barriers to such constructions are explained away as allegedly invalid. And it moves on to an ever-larger body of recent books on the theme of "autonomous agents," such as the impressive collection *Computational Theories of Interaction and Agency* by Philip E. Agre and Stanley J. Rosenschein. From these books—and many similar works—one could no doubt have whipped up an amusing introduction providing additional theoretical context to the ensuing conversations.

The reason we eventually decided against this really rather tempting introduction had something to do with the fact that it would make it difficult or impossible for us to carry out some opening moves which seemed very important. Anyway, introductions should carry at least some traces of humor or intellectual delight. And so we hit upon—as an inversion of and counterpoint to the previous style—a plan for an introductory game of confusion and vexation. This sort of postmodern introduction should tell the story of a Chilean student, who happens upon a collection of English tapes in the archive of a defunct institution. On these tapes, according to the Chilean student, are recordings from a seminar with one Professor Enriquez of the University of Santiago. Out of sheer curiosity, the student transcribes and translates this collection of tapes from the original English into Spanish, although his translation is never successfully completed. This Spanish manuscript forms the basis of a dissertation on the topic of "Contextual Problems in Machine Translation," which the Chilean student, now in the United States, the University of Chicago to be precise, submits to the Institute for Cognitive Science. A transcript of this machine translation manages, through neither chance nor necessity, to end up in an antiquarian bookshop in Heidelberg, where we discover it upon the occasion of an interview with Heinz von Foerster. And because even this meager little draft bears an uncanny resemblance to our Heidelberg conversations, we opt for a radically reconstructed German version of this typescript, which Heinz von Foerster ends up liking as much as if it were his own work.

Although it is not a German version of a Santiago student's English machine translation from a Spanish tape transcript of a conversation in English about the construction of the world and self-creation (rendered once again into English by an Austrian builder and an American nurse), the point of this introduction is that the gentle reader should gain an insight into how the text at hand came about. Nevertheless there was a whole cascade of steps between the transcription of tape recordings and the presentation of a conversation in book form. Such an introduction would probably have been amusing and would have afforded its readers a transitory diversion, yet in the end it did not seem to suit the book, and it would have omitted important information concerning the *dramatis personae* of those April days in Pescadero. And this brings us to a rather "autobiographical variation" on the introductory themes.

Such an introduction would foreground the self-doubts and editorial insecurities that we faced at certain points in our "Foersterian Self-Creation Project." This feeling was sharpest on the way to the Vienna airport, right before the departure to San Francisco, when the risks and contradictions of the projects came to the fore in the course of conversation: On one hand, we had already prepared a main thread for the seven days of conversation and expected a lively, open exchange, rich in surprises. From our opposite number, Heinz von Foerster, we expected elaborations upon key words for the topic of "nontrivial systems," and we now feared that, by means of a grandiose self-trivialization, we might have established the context too narrowly right from the start. On the other hand, to accept six days of spontaneous conversation—the seventh was given over to rest from the start—would probably meant drowning in circles, repetitions, and timeless confusions.

Although this form of introduction would have given us the opportunity to deploy our stock of devices borrowed from travelogues and bildungsroman, for precisely this reason it would, if carried out more fully, probably end up being too dramatic and theatrical. We thus ruled out this approach relatively quickly. The following variant was also seriously discussed: to focus on Heinz von Foerster right from the start, even in these prefaces. After all, it was he who had supported and encouraged the book project in several preliminary discussions—and who most considerately set up the necessary infrastructure for our dialogs—from picking us up at the San Francisco airport to booking rooms in the very peaceful Old Saw Mill Lodge, not far from the Foersters' home on Rattlesnake Hill. A final possibility, which knows only one version, was briefly aired: to choose a situation-independent "zero introduction" and to drop the editors' preface altogether.

Ultimately, the admonishing voice of our German publisher, who had accompanied our project in many supportive ways, ensured that we pursued none of these seven introductory threads—or any other as yet undiscovered introductory strands—in any greater detail. The manuscript was needed as soon as possible. And when looked at carefully, the current kaleidoscope of seven possible beginnings becomes an eighth form of introduction, which, in its own specific way, reflects the aims, doubts, moods and ambitions associated with this "trialogic book."

ALBERT MÜLLER AND KARL H. MÜLLER
Vienna, October 1997

FORETASTE OF AN AUTHOR
WITH TWO EDITORS

It just won't work. People come and say, "My dear Heinz, if we just put this neuron and that neuron together like in the brain, wouldn't we get a Heinz?" Then I say, "My dear friend, it doesn't work like that." "But these days we have all these new technologies, so why shouldn't it work?" Then I start playing the long game of showing why it does not work.

—HEINZ VON FOERSTER, in a conversation to prepare this book

Our knowledge forms an enormous system. And only within this system has a particular bit the value we give it.

—LUDWIG WITTGENSTEIN, *On Certainty*

And the Lord God formed man of the dust of the ground, & breathed into his nostrils the breath of life; and man became a liuing soule.

—GENESIS 2:7

We would like to approach your ideas and opinions step by step, moving from the fundamental questions of physics on the first day to the specific heuristics that characterise your style of thinking on the sixth and final day of conversation. On the seventh day, we will all have earned an unqualified rest.

There are many things I don't know much about. But you might say, that is exactly what we're interested in—a specific ignorance concerning this or that question. I could confess to that and not contribute anything very clever.

All of the topics of conversation have many points of connection with your works.

Then I will let you have the connecting points. If I don't have any ideas, I'll say, "I've got no idea," or "Let's look it up," or "Let's go for a walk." We can make an effort to keep a lively conversation going.

If we talk about this beginning before the beginning, that is, about the creation of this book, then right away there's a paradoxical constellation. You always write that the listener, not the speaker, determines the contents of a message or a sentence; the reader, rather than the writer, determines the contents of a message.[2] Now one might say that it's absolutely irrelevant what the writers, editors or designers do for a book, since it will all just be decided by the readers.

If you strike the word "just" from that sentence, then the sentence takes on a dynamic. It is not "just" the reader. It's the readers, the listeners who get a chance to become a creator, a genius, an inventor, an X, whatever that might be. What we always overlook is that when I open my mouth, something new comes out, even if it's only a silly or trivial remark.

I have two fundamental pedagogical principles, which I have taken from friends. The first fundamental principle is, "One can learn even from the dumbest." And the second fundamental principle is, "Let them die stupid." My university has these principles carved in blocks of marble. The façade is not embellished with busts of Aristotle, Kant, Leibniz, or Schrödinger, but these two inscriptions can be read: "One can learn even from the dumbest." This inscription is mainly—but not only—for the teachers. They can enter peacefully, without a guilty conscience. The second, "Let them die stupid," is mainly—but not only—for the students. Do we want to die stupid? If not, then we'd better listen up.

Those are my fundamental principles. I invite my conversational partner to make something for themselves out of the grunts and sibilants that I make with my mouth. It is a wonder that constantly goes unrecognized. I sit here and make these sounds with the air that blows across my vocal cords, with my mouth, in order to modulate something. And the other responds: "Yes, thank you very much," "Do you really love me?" That is actually unbelievable.

So I hope that the other will be able to make something of it. That's the game that we start. In a dance, first one person is leading, then the other. The music is there—they dance. Naturally I make my sounds in the hope that the other will be able to make something of them. And then I am so conceited that I even hope that he will make of them more or less what I hoped he would. He replies. Why does he reply? Because he too hopes that I understand him. He hopes that I will make of his sounds something that fits in with what I had hoped he would make of the sounds that I made. A very complicated sentence. But that is how I see the language game.

As a speaker, you learn from the listener's reaction to what you had said.

I have only *my* interpretation of the listener's reaction, not his reaction. I have no idea how he *has* reacted. I only see how I believe he has reacted. I would claim that misunderstandings, so-called misunderstandings, do not exist. There is only *one* understanding, namely, how I understood it. But maybe it is not what my counterpart had hoped I would understand. Then it is not a misunderstanding on my part. It's not as if I can misguess something. All that I have are his signs and what I have just heard.

In Wittgenstein there is the command: Say something and mean something else.³ And the point there is that although one can mean something different, that has no effect on the message, because that is determined by what is said.

That is something else, that is another game. That game goes like this: I spell out the word RED in capital letters, but the letters are green. Now we ask people, "What does this say?" One says "green," the other says, "red." With this RED written in green or GREEN written in red, we get at precisely what Wittgenstein wanted: Try to say something other than what you mean. The whole question, however, is—according to Heinz von Foerster—wrong. Or rather, it's not wrong, it's just not part of the problematics that I see in language. Since whatever you do with the air that blows across your vocal cords and makes a sound, I expect that the other—because this is the game—will try to make something of it. That is why we play language. And that whatever one meant doesn't actually exist at all. The game of language consists of both parties having the intention of making something, inventing something, constructing something out of the grunts and sibilants that the other produces. Now both are designers, making something of the utterances of the other.

NOTES ON THE TRANSLATION

The reader, not the writer, determines the meaning of a sentence.

What, then, do translators do? For readers in English, this trialog now has two additional participants, although these two are trying their best to be discrete, almost transparent. Try though we might, however, this is an impossible task: translation is not a trivial operation, and in true Foersterian fashion we translators have played an active role in determining the meanings of the sentences that follow.

In this trialog, the science cannot be separated from the person, and we strove to capture not just the participants' ideas but also their voices. The conversational dynamics are delicate: the joking, the bickering, the prickly wit of Heinz, the conciliatory efforts of Karl and Albert. Then, of course, there were all the puns.

This is a genre-defying project, a conversation that dismantles and ignores disciplinary divisions. Heinz, Albert, and Karl make it clear that computers, genetics, language, and *joie de vivre* really should be discussed together. This book is a conversation not just among three people but also among disciplines that are all too often not on speaking terms. It introduces a fantastic toolbox of mathematical and scientific concepts to the humanities. It also serves as a reminder that science and mathematics must not remove themselves from the dance of life.

Our translation itself has been a dialog. We have, in fact, almost translated the book twice, with Michael's first translation becoming the basis for Elinor's second translation, with much subsequent correspondence, debating, and fretting. The process has given us ample grounds for appreciating the magic of language and of mutual comprehension.

We began our translation in 2007 and finished nearly a year later. We returned to it—with surprise and delight—in 2011 for editing. The footnotes have been kindly transposed from the German by Albert Müller. We owe Albert Müller and Karl Müller many thanks for their encouragement and for their absolutely crucial assistance in bringing this translation to publication.

MICHAEL KASENBACHER AND ELINOR ROOKS, 2011

PREFACE TO THE AMERICAN EDITION

Heinz von Foerster invoked the metaphor of a fairy who through her magic powers accomplished a series of nearly impossible results, all of them contributing to the German edition of this book in the year 1997. Nevertheless, the most remarkable product of our faithful book fairy was the English translation of the German original. For this she needed to find persons with an impossible mix of qualities, ranging from a keen interest in Heinz von Foerster and from high competencies both in German and English to very rare features like an abiding interest in pursuing professional work even without any financial compensation. Our fairy godmother managed to identify Michael Kasenbacher and Elinor Rooks, probably the only persons with this particular combination of capabilities and skills, and provided them with sufficient strength to pursue the project of an English translation of our book against all odds and barriers. In combination these two translators managed to produce an astonishing result, namely a complete transfer from the frames and schemes of a German-Austrian book to the appropriate English contexts.

We can simply express our deep gratitude to both Elinor and Michael. Special thanks also go to Bruce Clarke, who accepted this volume in his series, and to Fordham University Press for including this book in its program.

ALBERT MÜLLER AND KARL H. MÜLLER
Vienna, August 2012

ABBREVIATIONS

CoC   Heinz von Foerster, ed., *Cybernetics of Cybernetics or The Control of Control and The Communication of Communication*, 2nd ed. (Minneapolis: Futures Systems, 1995)

OS   Heinz von Foerster, *Observing Systems*, 2nd ed. (Salinas, Calif.: Intersystems Publications, 1984)

PI   Ludwig Wittgenstein, *Philosophical Investigations*, electronic edition

TLP   Ludwig Wittgenstein, *Tractatus Logico-Philosophicus*, electronic edition

UU   Heinz von Foerster, *Understanding Understanding: Essays on Cybernetics and Cognition* (New York: Springer, 2003)

THE BEGINNING OF HEAVEN
AND EARTH HAS NO NAME

FIRST DAY

Building Blocks, Observers, Emergence, Trivial Machines

(i) Observations are not absolute but relative to an observer's point of view . . .

(ii) Observations affect the observed so as to obliterate the observer's hope for prediction. . . . In each and every moment I can decide who I am.

—HEINZ VON FOERSTER, *Understanding Understanding*

It is always by favor of Nature that one knows something.

—LUDWIG WITTGENSTEIN, *On Certainty*

In the beginning God created the Heauen, and the Earth. And the earth was without forme, and voyd, and darkenesse was vpon the face of the deepe: and the Spirit of God mooued vpon the face of the waters. And God said, Let there be light: and there was light. And God saw the light, that it was good: and God diuided the light from the darkenesse. And God called the light, Day, and the darkenesse he called Night: and the euening and the morning were the first day.

—GENESIS 1:1–5

In various places in your works, we find principles or aphorisms concerning beginnings. The following, for example, is an important proposition: The world or the environment contains no information. The world is as it is.[1] That means that observation or the observer is inseparably part of every beginning.

Where beginnings are concerned, we should see that we are sitting here, and every moment is always, always a beginning. "Everything is here and now," that's a magic saying for me, which I learned from my grandmother, Marie Lang.[2] "Everything is here and now." And so, for me, all the problems of history, the distant past, the beginnings of the universe are here and now, the stories about them are always constructed here and now. Every time we talk about them, these stories take on a different form, are in a different context; they change when I talk to Albert or to Karl or when I talk to both. The here and the now—for me this is a central point—is the beginning of every beginning.

Therefore the here and now is also always a new creation, a kind of "genesis." The moment in which I do what I do is always new; it was never there before. Nothing was before, because what was is as I think it was before. That means *I* tell how it was. And that results in an extraordinary problem of responsibility. I can still remember the big motto in the Stanford School of Journalism that said, "Tell it like it is." When, to my horror, I saw that motto, I walked in there and said, "Listen, ladies and gentlemen, it is as you tell it, and that's why you're responsible for the 'it.' Because you tell 'it,' it 'is' as you tell it. You can't say how it 'is'—no one knows how it 'is.' And when it 'was,' no one can reconstruct *how* it was."

Here someone might object, "But we can substantiate this or that; we have witnesses." Then listen to their accounts again a few days after the so-called incident: Then the car was driving fast, turning left, turning right, a lady was at the door, but no, it was a man, a child. . . . Everyone tells a different story. A good movie in this vein, by the way, is *Rashomon*.[3] A woman claims that she has been raped and that her husband was killed by robbers. And she manages to gather witnesses to the murder, and then they tell their stories. Each has a different story—and they all fit. *Rashomon* is my idea of "Tell it like it is" versus "It is as you tell it."

Of course, here we are already right in the middle of the so-called "problem of the observer" and in the middle of the beginning of our beginning. The beginning of creation must also be observed.

The story of the creation, yes, of course.

In the case of the biblical creation story, the narrative works wonderfully, so that the creation and the observation are one in the same, in that "Light" is proclaimed, and it comes into being. At least that's what we read in the Bible.

Before this creation there had to be some kind of ur-Creation in which God said, "Let there be sight," and "there was light." One must first be able to see in order for light to come into existence. Because light is not just there.

Actually now we have to go back to the ur-ur-Creation. How can one see something before the beginning?

Well, I'm going to play God now. First God/Heinz would have to invent sight so that light can be welcomed in. God/Heinz would invent ears so that sound can be heard, for there is no sound. There are only molecules— apparently—that move at a terrible speed. It's only when they first happen to knock against the eardrum that you hear something. There is no music, just as there is no coherent light. Electromagnetic waves, the physicists claim, are swinging through space. But that has nothing to do with light. I can see light only when I have something that produces light perception when confronted with electromagnetic waves. Sight comes before light. We can carry out experiments to show that all colors are an invention of the brain, an invention of the eye. We can do experiments and show that no colors exist at all—and yet we "see" colors. But I get what you're driving at. Dear Heinz, you're thinking, now you're fleeing into excuses: Let there be sight, and there was light. Then where does sight come from?

We are playing the game of beginnings, and another problem that poses itself here is that of the many different beginnings. The creation story of Genesis is one beginning. In other cultures we find other beginnings, for example, the story that the earth rests on a turtle, and so forth.

Or that the world came into being through the Big Bang.

These many beginnings or these many stories of the beginning refer us on one hand to the problem of observation and the many contexts and cultures that generate these stories of the beginning. On the other hand, there is the question

of whether such stories are meaningful at all. Is it possible to connect these sto-
ries of the beginning with decidable questions?

These stories belong to basically undecidable questions. They are mostly
a game to find out who the other is: "How did the universe come into
being?" If I hear the answer "Big Bang," I say, "Thank you, that's astrophysi-
cist talk"; or, "Every child knows that God created the world in seven
days!"—then I know that's a good Catholic, of not quite the latest fashion;
or someone tells the story of the turtle-tower; or someone is Heinz von
Foerster and would then ask, "How would *you* say the world came into
being? When you have told me that, then I will know who *you* are." It
deals with a basically unanswerable question: no one was there, no witness present. And suppose there was a witness; maybe he would lie to us?
We don't know whether he's telling the truth. We have no opportunity
for comparison. We would have had to see it ourselves—and we didn't, or
else we wouldn't be asking.

One immediately starts turning logical somersaults when one asks
about the beginning of the beginning. I want to draw attention to that.
The question of the beginning is one of the fundamentally unanswerable,
undecidable questions, and from the way it is answered all I can learn is to
which cultural milieu, which language milieu, which personal milieu,
which belief milieu the person with whom I am discussing the question
belongs. Having said that, it's clear that I am speaking about myself and
only about myself. A reader reading that could say, "That Heinz is a cow-
ard, that Heinz is lazy and cozy. He doesn't know what he's talking about."
Such a reaction would be entirely appropriate because it is the listener who
interprets an utterance. I can only hope that he or she knows what I'm
talking about.

We must decide all undecidable questions for ourselves.[4] *What decision have*
you made for yourself in the game of beginnings?

I have decided to ask my friends how they see the beginning and to get to
know how they react to this. I know it is an unanswerable question. I don't
need an answer. And so I ask my friends. And this is how they answer the
question: "Who are you?" I move the creation-of-the-world question onto
the who-are-you question, which is answered in the answering of the
world-question.

You surely know the important book *Laws of Form* by George Spencer Brown. On one of its first pages, this book has a motto in Chinese characters. I once asked a student of Chinese origin what it means, and he translated it for me: "The beginning of heaven and earth has no name." An incredible proposition, don't you think?[5]

Astrophysicists and physicists could now explain that, with the Big Bang, observations, conjectures, and counter-conjectures may be found so that the matter of the Big Bang will become a decidable question.

Why astrophysicists consider it decidable, I don't know. I know it's undecidable. I'll draw a comparison. The situation is like in chess: You choose a move, and that is the moment when the undecidable question becomes a decidable one. You're saying, "We want to play a certain game now; it's called astrophysics." What are the rules? We make observations with telescopes, we build space telescopes, we know spectroscopy. We know what Doppler wrote about wave movements, frequency movements, and so on. Within these rules we want to find out how the world came into being. Thus we come to certain conclusions. That means that in the matter of beginnings, the unanswerable is a question of which game I should play. And if we all decide to play the game of astrophysics or physics or chess or checkers or backgammon, then the undecidable first decision is made. Because until then it was basically undecidable which game I should play—this, that or the other—maybe arithmetic, mathematics, or the numerical system.

It is *my* decision to call numbers mathematical objects that fulfill such and such conditions. Sometimes they do, and sometimes they don't. Then what happens? Then I invent a new ensemble with numbers. If, for example, you want to take five away from three, then you have a problem: such numbers don't exist in everyday life. I can't put five books on the shelf if there are only three here on the table. In everyday life you can't take five from three. But then someone else says: Look here, of course you can take five from three if you have numbers that mark out these special operations. If you put a little horizontal line before the number that you've already taken away, then you have a number with a horizontal line before it. That's called "minus" and you get −2. You can perform operations with that as if it were a +2, as long as you incorporate these additional rules. It was an invention to solve this problem this way. And so through this elegant invention a basically unanswerable question suddenly became answerable. A new game of negative numbers was born.

In Karl Popper we find a distinction between discovery and invention. For Popper, it was an invention that brought numbers into the world. But after that everything else is a discovery, whether it be negative numbers or irrational numbers and so on—once the basic numbers, numerical system, series of numbers are fixed, that means every further development is a discovery in the remarkable World Three.[6]

The first part is quite right. But negative numbers are no discovery; they're pure invention. It's a discovery if there are positive numbers: two times two is four. I didn't know that before—but it was always already there. We want to discover: what is two times two? Hocus-pocus, I draw back the curtain and there stands four and not three. Three minus five I can't discover. That's not provided for within my rules—therefore I must invent something. If the framework, the rules of the game are in place, then everything else is properly a discovery. But if I come across something unsolvable and I still find a solution, then it was just an invention.

Would irrational numbers or differential calculus be inventions in this sense?

All inventions, constructions, inventions. It must deal with a problem that's unsolvable: the square root of -1. What do I do now? Invent imaginary numbers—they're even called imaginary numbers.

With regard to cosmic beginnings, an important question about matter is that of penultimate and ultimate building blocks. In the twentieth century there seemed to be a game of Russian nesting dolls—first atoms, then the atomic nucleus, then quarks, then conjectures toward a theory of "strings."[7] Scientists have generated a cascade of ever-smaller units. Is this whole search for the ultimate building blocks even meaningful?

That depends on who you are. For some people it's a very meaningful search because they're in the building blocks trade. If you're in the building blocks retail trade, you will always look for the smallest building blocks. If you're not active in the building blocks trade, then you'll find these building blocks to be a comical aberration on the part of your colleagues. There are so many different ways to build. First we have the building blocks idea. I take bricks, fit them together, put them here.

Another method is to say as follows: Now I want to build something that will fulfill a certain function, for example, that shelters me from the wind. Maybe then I don't need any building blocks at all; maybe I can bore a hole in a tree and creep inside—and I'm sheltered from the wind then, too. It all depends on what I would like to have at the end. If I want to look for building blocks just to have building blocks then, as I said, I am just working in the building blocks trade. Otherwise it is left open as to how I will handle the problem I now have before me. I want to shelter from the wind, I would like to eat, I would like to play this or this, I would like to explain such and such. Do I need building blocks for that? No, only someone who is acting in building block mode, who needs building blocks to brace other building blocks. If you don't operate in the building blocks industry, you can do completely different things. Take physics or mathematics, which offer the most elegant and beautiful solutions for this. For example, everyone makes fun of *Candide*; in it Pangloss claims that we live in the best of all possible worlds.[8]

It's interesting to examine this best of all possible worlds story formally. Mathematicians have also taken an interest in it—and have given it its own area, called variational calculus. An example from physics: There's a bowl in front of you. On top of the bowl there is a little ball, which you allow to roll. Which way will the ball go?

The best possible direction for the ball is the direction it takes. That's an interesting result, one in which, incidentally, causality is nowhere to be found. Here all that is important are the marginal conditions of the universe in which something takes place, by means of which a certain quantity is minimized or maximized. What we have here is, one might say, a systematic idea that assumes a totality and doesn't actually need the ball. I won't know whether it rolls left or right or whatever—it's absolutely not necessary for the result. This idea of maximizations, minimizations, and optimizations of systems is a way of thinking that is constructed differently from the building-blocks form. Here building blocks do not appear.

What I'm saying with this is that there are forms of thinking that don't require the concept of building blocks. If, however, you are in the business of small and smallest building blocks, then you are constantly searching for new tiny building blocks, even tinier building blocks, the tiniest of tiny building blocks. And our colleague Pangloss, who is always being made fun of, is constantly employed in physics. The problems of dynamic equilibrium, which Prigogine dealt with in the last decades,[9] were already recognized in the eighteenth century. Many things are therefore physically optimal because they are in dynamic equilibrium, because something has been optimized and a path or a way has been found that is subject to certain constraints and in which something has been minimized or optimized.

In Leibniz's original concept of the best of all possible worlds,[10] the rationale was as follows: God could have created a world that was fundamentally more organized than the one we have. On the other hand, he could have created a fundamentally more chaotic universe. We have, therefore, a world in which order and chaos are at a maximum. That is a central point, a criteria, for why this world was chosen and created in this way—and no other. Here we also find the vision of order and freedom jointly maximized. What, from your perspective, is maximized or minimized with regard to our question of cosmic beginnings?

One might use action as an example. Action is a time-integral over energy—and I am now using the principle of least action just as it is now used in physics. The word action is a physical dimension. If you take the physical dimension of energy and take the integral of that and calculate by means of the minimizing calculation how the processes are running, through which conditions this system is passing, then through this the energy will thus be minimized. And differential calculus can play this game optimally. It offers a solution, showing under which conditions certain processes will be maximized or minimized. I can find extremal values.

That's a very elegant method, which I liked very much when I was a young man. When I was dealing with physical problems, I was—much to the horror of my teachers—always inclined to try to reduce these problems to the principle of least action and then carry out the so-called calculus of variations. So one writes that delta—the change in a process—equals zero; thus is the effect minimized. I always greatly enjoyed this principle. The ideas of Bernoulli or Leibniz, the Voltairian figure of Pangloss, always seemed ingenious to me. One need not be concerned with irrelevant details at all. One claims that the world is constructed in such a way that a maximum or minimum is the result.

If someone were to say that least action shouldn't be foregrounded, but rather that the magnitude of complexity and order should be put center stage—what would your reaction be?

If they're happy with that, then they should just go on working with it. I think that no longer has anything to do with the issue at hand, it has to do with Mr. Jones, who likes complexity. If someone is dumb enough that he wants to deal with complexity, he's going to stay that way. But: "One can learn even from the dumbest," or, "Let them die stupid."

There is, for example, the so-called anthropic principle.[11] In the weak form, it states that the initial conditions and the choice of constants, natural constants, are so ordered that life, diversity, and ultimately humanity could develop. Had these constants been different, things would not have turned out this way.

Very funny. A nice idea. If the constants had been different, it would very probably not have resulted in humans but in elephants or something else entirely. But: Everything is here and now. *Now* do you want to explain the anthropic principle to me? Thank you very much, if you want to explain that to me here and now, then I know something about you. You would like the world to be such that you had to result. I can already see where you want to end up. You want to force the world to turn out as it is; and maybe, if I'm being a little vicious, I'll say: Would you like to use this as an excuse, so that whatever happens tomorrow had to be, and you couldn't have had any influence on that tomorrow? Do you want to use the anthropic principle to avoid responsibility, because, as with fate, nothing could have been otherwise? Do you think that fate has made you that way? Another form of Kismet? An anthropic principle? It could not have been otherwise than that you are here—the world has constants and so forth, doesn't it?

Again, it comes back to how you, as a human being, want to represent things, in order to make something of yourself or the world or others. And that is left up to each of us, each in our own way. The anthropic principle cannot be proven because it is always already proven, we humans are already here.

Either one is in the building blocks trade, or we are in the order industry, also a business, and as regards the question of the final building block there is no end in sight. In twenty years—or maybe even five—someone who is interested in it will receive different answers than they would today. A closely connected question is: What holds these building blocks together fundamentally? There's this idea that the universe is a grandiose machine, describable by fundamental laws.

Exactly, and a grandiose *trivial* machine, to be precise. But if you try to build the world as a trivial machine, you leave yourself out. That is, you're not a part of that world. For me this description of a world without the people who describe it is boring. I'd much rather ask: How does that happen? How could I build a world—if one does want to build it—in which I too have a place, in which Karl Müller and Albert Müller will also appear?

Clearly, Karl and Albert are not trivial machines. And one can prove that very easily by asking them, "Do you think you're a trivial machine?" And each will say, "No, I'm not a trivial machine." If I do not wish to incorporate the citizens of the universe, this naked universe is going to be very dull and monotonous.

Naturally, it can be played as a game; that's certainly fascinating. But without my personal relationships, which I would like to maintain, with my wife or with you and my friends, it becomes empty and boring; and there is no opportunity for me to watch a dance, play in a jazz band, to look at pictures, to be pleased or annoyed by Picasso. None of these elements could be found. Of course, I do find this empty world amusing as a game, as an intellectual challenge, as a case study in the smallest units, as a somersault in the *n*th dimension.

That's all very amusing, but the whole point is that I couldn't enjoy Picasso, or delight in Grete Wiesenthal's dance,[12] or whatever. I want to embrace them all in my worldview, and here I must leave the people in the building block trade. Why do I have to leave them? The strange thing is that if one has bricks, then one will always build houses that people can immediately tell are made of bricks. If I were to build houses out of chewing gum, the houses would look completely different. The building blocks that I use to build something influence very strongly what is built. A building is always going to be an expression of that from which it is constructed. And so I begin by leaving out the building blocks, because first of all I want to be a co-resident, a part-taker, a part-giver of the whole.

If you liked, you could ask me, "So Heinz, you want to be a building block yourself?" Well, okay, perhaps I am one of the building blocks, why not? You see what I mean. Let's see what happens if I use a Karl Müller, an Albert Müller, a Mai von Foerster, and a Cornelia Bessie as the building blocks of our world. Now, do you have me where you want me?

Let's talk once more about the question of the construction of the world: In a beautiful article, John Wheeler formulates a heuristics of certain substantive noes and five yeses. The first "no" concerns precisely this idea of building blocks: "No turtle-towers," as in: These cascades of turtles should be avoided; the idea of building blocks is not fruitful. The second "no" is "No laws."[13]

"No laws," I like that. Let's take for example the laws of nature. What happens if the laws of nature are not obeyed? For example, according to the Newtonian laws of nature, the planet Mercury is supposed to move in a certain way. The planet Mercury, however, does not move as Newton prescribed it ought to. It moves differently. Mercury's behavior could not be predicted until Einstein's natural laws—Newton couldn't do that. Now, if a law of nature is not obeyed, the legislator will be arrested. If a human law is not obeyed, however, the one who broke the law will be arrested. My proposal regarding this double law breaking is as follows: whenever any given law is broken, arrest the legislator and say, "Now make better laws." For once let's follow the natural philosophers, who always arrest the legislator if something goes wrong.

The third "no" is "No continuums." All natural laws are to be described by discrete models and not continuous ones. Continuous models, differential equations, prove inadequate as descriptions.

Continuous models are a perverse distortion of what is actually the case. Let's look at the conception of a continuum. I can say, for example, that numbers follow each other, so I have one, two, three, four, five. Now I can find a point between one and two on the number line. Can I place yet another point between one and the point I've just found? Now we are entering the game. The game is called: I'll never stop finding points between points. Do we want to play this game? Someone else says, "No, I'm not playing that, it's too stupid for me." The first one says, "You're just too stupid to get this game." So now both have generated their different rules. One plays at endlessly finding a point between two points. The other, however, is convinced that this game is too monotonous, too unrealistic, or whatever else. "I have one, two, three, I can live with that just fine. After all there's only one dollar or two dollars. I do have cents as well, but with your game I'd *only* have cents (and would be a poor man); and furthermore, cents are the smallest unit."

Let's call such rules a grammar: Which grammar should I use so that I can get into semantics? That's the point I want to drive at. I want to find a distinction between the grammars, the formalisms with which I wish to connect my semantic units, my "bubbles." Does something help me in this, not help me, or does it disturb me in other ways? I always leave this question open, I would always like to start again. Every moment is a beginning, an origin. Now I will look at it like this, later like that.

Wheeler continues: "No space, no time."

Is there space? Is there time? The whole problematic, the questions about space and time, the discussions or theories, all have to do with the existential operator "is"—with three letters in German, two in English. To be or not to be? In "is" lies the trick or the trap, into which one falls easily. If one looks at the clock, there "is" time. But if someone forgets to show up on a date, there "is" no time. If one gets on the tram and travels from A to B, there "is" space. Anyway, what "is" this "is"? Now, let's play "Space." Then we can quarrel about whether it "is" or "is" not. But then we have generated space as something we can talk about, whether it is or not. But the "Is" keeps cropping up, and nobody deals with it.

I want to draw attention to the "Is." I hope you understand my shifting from the question of space to language—it "is" not a question of space, it becomes the question of "Is." "Is" there space or not? It comes down to how we want it. If we want to have space, then we take the train and go from A to B. If we want to have time, we look at our watch and say, I'll meet Karl Müller at eleven. And if I am late, I'll say, I'm sorry but I was unfortunately delayed. All these situations occur, then. But one can also leave them out. You can say that has nothing to do with it, we are sitting here, here and now. And everything else is just unreconstructable past and unknown horizons, which I refer to with noises, with my grunts and sibilants.

For Wheeler—as for you—observation is a very important element that cannot be eliminated: No question, no answer. If you turn this proposition round, one must already have certain answers in order to pose questions or make observations.

For example, there's a book called *What Is the Name of This Book?* And if one asks, "What is the name of this book?" the answer is *What Is the Name of This Book?* That belongs to a very particular class of phenomenon in which questions and answers are the same. This book title is a wonderful indicator of how little questions can be separated from answers. Ludwig Wittgenstein is even pithier: "What is a question?"[14] That's the best way to annoy someone: What is a question?

Another important piece of advice from Wheeler lies in the remark: More is different.

That speaks to the important question of quality and quantity. Can quantity become quality? We so often hear, well, quantity is a naïve variant, a trivialization of quality: "You only ever see the quantitative!" There's that word, "only." The number three is actually fundamentally different from two. If one sees three, four and five, one also sees the cognitive functions of such so-called numbers, which is, by the way, connected very closely with the problem of counting. Counting is very often seen as a temporary or temporal activity. First 1, then 2, then 3, then 4, and so on. I've never tried, but I would need maybe an hour to count up to 1,000 like this. In any case, counting costs me time.

However, counting can be seen as a qualitative matter, namely as four-ness, ten-ness, or eleven-ness. To illustrate: If I take a pair of dice, with which one might play Monopoly, for example, I look at them and "know" it's a six. I look again and "know" immediately that it's a three. Here the configuration of three-ness, two-ness, or six-ness presents itself to us. In the sense of this distinction, the number becomes at one point part of a continuum that demands time of us; another time the number belongs to a configuration or a relation, it establishes the connection between looker and looked-at.

The idea of seeing patterns, configurations—"pattern recognition" is a very popular field of research these days—leads us, it seems to me, to a dead-end. If we look somewhere, we can of course "create" patterns—and therefore we have also "seen" them. In this case, then, there is "three" or "six" on the dice. But if I *cannot* see that these points are arranged in such a way, then the six or the three suddenly *isn't* there. The six only comes into being through my making six into six—and then the six is there. It has to do with emergence, it has to do with "always seeing something new." I would also say that six is not discovered, but that six is created when I look at it. Now of course we have learned very well who created six—we know that by now. Just a glance and I spontaneously say, "A six."

But that also means that the difference between quality, qualitative, and quantity is only a difference of organization.

Exactly, that's why I went into all that.

Now we still only have the first part of an answer to the topic "More is different." Here too we could start from a quantitative organization, which if we expand it could . . .

It could become something different, and just because the configuration is different: seven-ness does not look like six-ness; that's the whole point of "More is different." Seven-ness is something different from six-ness. 25,000-ness is different from 21,000-ness. Apparently these differences disappear when we're dealing with big numbers. But these differences again become apparent when dealing with very, very big numbers, such as the number of configurations in the universe, which comes—no matter how you guess at it—to 10 to the power of some very, *very* big number. If we're dealing with these really, *really* big numbers, we actually lose our connection to them. Then there are forms, forms we can't even look at. Here we take the form much more as a quality and work with these other forms qualitatively.

But even going from six-ness to seven-ness and to ten-ness there are jumps, sometimes big ones and sometimes smaller ones.

Of course, and new differences time and again. And even a small difference can be immense. Take six and seven—or three and four. It is incredibly multifaceted what happens when I make four out of three, everything that I must undertake so that four emerges from three.

Quantitatively thinking, one simply has to add one, and here quantitative calculation displays its well-known advantages.

I'm not at all saying that the quantitative method is a swindle. I simply maintain that there are disadvantages for a culture if it concentrates on only one view of numbers, of numerality, while dismissing the others as inferior. Because the two areas, the qualitative and the quantitative perspectives, should be seen as complementary. One needs the other in order to form a "totality." It's only when I can swim in one world as well as the other can I really swim, and most of all, I will have more fun when I swim. And someone won't be called an "idiot" just because he counts with beads or perceives "three-ness."

From this interplay between the quantitative and the qualitative, an important restriction arises, indicated by the word "emergence." Currently emergence can mean many things; one meaning is that over the course of time something comes about which could not have been foreseen at the beginning.

When did emergence emerge? As far as I know, emergence first emerged only a few years ago—suddenly there was this buzzword. Now I want to examine the problem of emergence in a rather mean-spirited way: the moment such buzzwords appear I immediately put on my political hat and ask myself, "Who invented that? Why did they invent it? And what happens as a result of this buzzword?"

In the case of emergence—we are now in the middle of our game of beginnings—there are at least two problems. The first consists in the limits of explainability and predictability at a given moment. Conversely, what is interesting about emergence is that over the course of time new phenomena come along that— although the "building blocks" already existed—were nevertheless not predictable. Are there—in your opinion—phenomena for which all the components are existent, but their recombination remains nonetheless unpredictable?

Here, too, we require the important distinction between trivial and non-trivial systems. First let's take a long step back. Let's look at an operative unit and claim that it carries out certain tasks, has a specific function or is able to perform diverse transformations with an operator. Now there are operators that are simply fixed and finished—and always do the same thing. Such an operator could be a multiplier by 2. You give it 4 and out comes 8. You give it 5, you get 10; you try with 1 and the operator responds with 2, and so forth. In this case you have what I call a trivial operator. Why do I call it "trivial," and why do I call such machines "trivial"? Because analytically they pose a trivial problem. I would like to know what this operator is doing and so try the operator. I give it ten or twenty goes; it always comes out the same; 2 becomes 4, 5 becomes 10. This machine is analytically determinable. Synthetically defined, analytically determinable. But now I take an operator that, once it has carried out an operation, changes its own operation—we might also say, that changes its inner condition. After a multiplication by 2 it becomes a multiplier by 3. And if it was with 3, it goes on to multiplying by 5 or something similar. And now you let this operator "run" and ask a friend, "Could you find out for me how this machine functions?" We seem to be dealing with a simple analytical problem—and yet it turns out that even with the most elementary nontrivial operators these analytical problems are unsolvable.

There are two kinds of insolvability. First, something is, as one says in computer science, transcomputational. "Transcomputational" means that the number of possible interpretations of this machine is just so great that the age of the universe would not be enough time to calculate them all. Second, one can build a system that is nonanalyzable on principle. Now, if I look at the world and say, I'm going to cut a piece out of the world that I would like to observe, and then I would like to say how it will proceed or how it operates, then I formulate the analytical problem for units or phenomena that are cut out of the world. From the beginning I must ask: Do I cut out systems that are trivial, or ones that are not trivial? Once I've recognized that I am dealing with a nontrivial system— because after every operation the operation of the system changes—then right away I can say that this system is unpredictable, that is, that it is not explicable through analysis.

Once I spent a month playing the following game with the local radio station. Every morning from 8:05 to 8:07, the local station broadcasts the economic news. In almost every broadcast I heard, "Unpredictably, the following has happened . . ." As I counted it, in 90 percent of the broadcasts, we heard something like this: "Such and such has happened unexpectedly," "This happened unforeseeably," "There has been an unforeseen development in the economic indicators." Isn't that the way? Don't you hear economic news in which every sentence starts with "unforeseen"? Don't you hear them talking about economic indicators that are fundamentally worthless? They're only valuable for those people who say, "I am going to predict for you how the system will work." Here is the nontrivial system, and here is the clairvoyant. In the meantime, these clairvoyants or "obscurevoyants" get paid a lot of money because they maintain that they can *correctly* predict the future. That they are constantly wrong doesn't matter as long as you just believe that they are soothsayers. That's certainly a mean thing to say, but actually it's all very plain and simple: As soon as I isolate a system in which I suspect that the result of an operation will work back on the next operation, in any form whatsoever, then this system is no longer predictable.

Another meaning of "emergence" is that which says that atoms have no colors as such, meaning that the color characteristic becomes visible only within a certain spectrum of magnitude. In this sense there are all kinds of emergent qualities, such as "consciousness," for example: a single nerve cell has none, but we ascribe "consciousness" to an assemblage of such cells.

How can someone claim that? He probably wants to see it that way. He poses the problem to himself by saying: "The single nerve cell has no consciousness. What about two? How many cells make consciousness?" Then I say, "Terrific, what are your thoughts on that?"—"27!"—"Fabulous, great." Someone else says, "Nonsense, 27 is far too few. I need 22 times 10 to the twelfth power." Aha, wonderful, a clever man with a head for big numbers. But that's just the beginning of the game of confusion: "Now please tell me: what is consciousness?" And then you've got to listen to what all these people have to say about consciousness. At this point all we know, at best, is who each person is. Because no one can tell me what "consciousness" is. I can of course look it up in a dictionary and then we can all discuss whether we're happy with this definition or if another would explain things somewhat better.

My position is as follows: emergence is a culture-bound cognitive process. I live in a certain culture in which I can see that certain phenomena "emerge" and others do not. We find emergence where something new comes into being, where one or some or many people or we ourselves suddenly see something differently. The moment I see something differently, something new is there. You can call this "emergence" and thereby remain generally understood. However one could also bring the perspective of self-organization into play: This self-organizing system has swung into new eigenvalues that it did not previously possess. And with this there are new insights to observe, something has "emerged." But not there—no, here, inside of *me* something is newly configured, and *I* see it as a new understanding. Emergence is *my* ability to see newly.

In many cultures this seeing of new configurations, of new ways of seeing, is supported; in some cultural contexts it is suppressed. I'm reminded of the following story: If Leonardo had a student, he would put a block of marble in front of them and ask, "How many faces or bodies do you see in this block of marble? How many figures, how many bodies? How many animals, how many objects do you see in this piece of marble?" "I only see marble!"—"Oh dear, I can't take this one." Another candidate: "By God, Achilles is fighting Hector!" If you are in the position of being able to see the Trojan War or other somewhat nicer episodes in marble or in the grain of wood, then a great deal "emerges."

Behind the sometimes-trivial variants of the concept of emergence stands an old assumption: The whole is more than the sum of its parts.

I would add the following the correction to this principle. You require an additional measurement function for this: The measure of the sum of the parts is greater than the sum of the measures of the parts. One is the measure of the sum; the other is the sum of the measures. Take, for example, the measurement function "to square," which makes this immediately apparent. I have two parts, one is a, the other b. Now I have the measure of the sum of the parts. What does that look like? $a+b$ as the sum of the parts squared, $(a+b)^2$ gives us $a^2+2ab+b^2$. Now I need the sum of the measures of the parts, and with this I have the measure of a $(=a^2)$ and the measure of b $(=b^2)$: a^2+b^2. Now I claim that the measure of the sums of the parts is greater than the sum of the measures of the parts and state that: a^2+b^2+2ab is greater than a^2+b^2. So the measure of the sum is greater than the sum of the measures. Why? a and b squared already have a relation together. When you have grasped this point, then the first question is: which measures do we want to use if we are speaking about "parts" and the "whole"? What is the definitive measure for me when it comes to the parts and the sum of the parts? As soon as one has decided that, one can immediately tell where the differences between "parts" and the "whole" lie. My assertion about cooperation as a superadditive configuration is founded on the fact that with cooperation the measure of the sum of the parts is greater. A competitive configuration is a subadditive configuration because its end result is less than in the other case.

You can conduct the conversation on one level, namely positing emergence as a sacred cow and saying, "Where are the udders? Where are her horns?" For once I would like to carve up the sacred cow and then see what we can start to do with the cow.

A butchered cow, however, no longer gives milk.

Sure, but you've got her parts.

It depends how the joints are connected. Earlier you emphasized that one ought to mind the operators.

Exactly! To mind what one would like to see in the operators.

At the moment there is an important argument concerning such operators and the limits of our knowledge, or rather, its predictability. That's Popper's argument— that at the present time we cannot predict our future knowledge.[15]

That's so trivial for me, since we're clearly dealing with nontrivial systems, which are absolutely not predictable. We don't need Popper's "knowledge" for that.

According to Foerster, then, it is not only knowledge that displays this strange property but many other systems too. It depends on the particular operator.

I would even say that nothing is predictable. All systems that we isolate from the universe are nontrivial systems. Our hope that they are trivial is, looked at carefully, a naïve hope. Even that best of cars, the Rolls-Royce, which is sold as a guaranteed and everlasting trivial machine, can break down one day. All the fabulous trivial machines that come with warranties—if this doesn't work, give the machine back to me—and which therefore cost thousands of dollars, are ultimately nontrivial. If the salesperson says to you, "Guaranteed trivial," he or she is a scoundrel, an idiot, or both.

This position seems "to emerge" for the first time in our conversation in this rather pointed form. Counterexamples would be, of course, planetary orbits, the orbit of the moon, and past and future eclipses. Those seem to remain stable over very long periods of time. The clairvoyants have withdrawn their interpretations of eclipses.

The planetary system is a little like a Rolls-Royce—it almost always works. Poincaré has already pointed out, however, that this planetary constellation is not a two-body problem but a three-body problem. And so, can we solve it? In principle, no. If one is a cosmic mayfly like man, then of course our solar system works in a trivial way. Then it's possible to maintain the apparent triviality. Moreover, just being able to say "sun" and "moon" already creates a certain substantial invariance. An important invariance of the moon lies in my being able to call her "Moon," and in her case that is actually quite incredible. But these astronomical invariances are only possible because we are cosmic mayflies.

The landscape of devices of our everyday life—car, washing machine, TV—is actually conceived of, apart from in slapstick movies, as a trivial machine.

It's always conceived of as a trivial machine, and that is why the radio journalist, when he wants to describe a trivial economy, says, "Unexpectedly, today the index has gone up or down." He is constantly admitting that someone was wrong, but people have learned to live with these mistakes. They say, "Aha, if I buy today then I change the system. How much do I have to buy to change the system so that I can be back in business?" In this way, second-order triviality emerges.

A single nerve cell is a trivial machine?

Doctors would claim that. I would counter that it is *nontrivial.*

Let us take Pitts's and McCulloch's model of the nerve cell.[16]

However fruitful it has proved for researchers, this model consists only of trivial elements. Thus, the Pitts-McCulloch system is a trivial system. The beauty of the Pitts-McCulloch system lies in what John von Neumann developed out of it. John von Neumann has shown that the Pitts-McCulloch system is equivalent to a Turing machine.[17] And further, he has seen a connection to language. Here we can insert the logical function X or Y—X: the weather is nice, Y: it is raining—and then express logical statements with it, a first-order logical calculus. That means that everything I can express in language—not in function calculus but in representation calculus—I can describe with a Pitts-McCulloch net. In my opinion, that is an important insight of Neumann's, which, by the way, I like very much. This brings us to the question of why we can say of certain machines or of a human economy or of society that we cannot "picture" them. It's because they're nontrivial. All the elementary particles of which these systems are constructed are nontrivial in nature, meaning that once carried out, an operation changes the operating style, the operating form of the machine.

As soon as I find nontrivial elements in a system it becomes a nontrivial system. That is an important theorem. Now, one can take such theorems into consideration, or one can also ignore them and set them aside. Then one can continue to have all kinds of really funny conversations. When I say that we can have funny conversations and amuse ourselves, I'd like to illustrate briefly the ways in which people talk about the future of economic development. It's really very amusing to listen closely: "You're an optimist?"—"Oh dear."—"You look like a pessimist. It can't be all that bad." That's how these conversations go, and it works, too: People talk, they play, they buy shares, they don't buy them, a party game like Monopoly develops, only incomparably more expensive, more extravagant. But to talk about tomorrow's economy today, *it doesn't work*. One should always keep sight of that. I don't claim that one shouldn't talk about it. But that talk belongs to another category, namely entertainment.

One more question about nerve cells. If it's clear that the Pitts-McCulloch nerve cell is a trivial machine, wherein lies the difference from the Foersterian nerve cell, which is nontrivial?

The difference between the two cells is as follows: In a Pitts-McCulloch cell, the nerve cell responds and fires whenever a stimulating impulse hits the nerve cell. The Foerster cell does the following: One time an impulse reaches the cell, it fires; the next time the impulse reaches the cell and it says, "Not enough." According to Pitts and McCulloch, every nerve cell has a threshold. If an impulse crosses the threshold, these cells fire again. My cell does the following: One time it fires, and now the threshold is raised a notch—and the next time it no longer fires.

With this, its inner condition has changed.

Yes, the operation has changed the inner condition, the threshold is higher now. In McCulloch-Pitts nets the threshold stays the same. The only way the threshold is changed is by accepting an inhibitory impulse. That means that I can now inhibit the cell through another loop so that if the impulse comes the cell does fire. But then word comes from elsewhere: "Be the sort of cell that won't fire anymore." Conversely, the

Foerster cell raises the threshold itself, and in such a way that I cannot determine it, so that the cell remains for me unexplored and unexplorable, a nontrivial cell. Pitts-McCulloch cells, on the other hand, are all trivial cells that are regulated from the outside.

Looking at it rather more generally, then, at one extreme we have Laplace's daemon, who tells us that the universe is built as a trivial machine: "Give me any point, and I will tell you the future and the past." At the other extreme is Foerster's daemon, saying, "Everything is nontrivial. Give me any point—and here I will stay, I can go no further." Let's suppose these two daemons were to connect and create a new daemon. We'll call it "Laster's daemon," from Laplace and Foerster, and it has the wisdom to say, "Treat this system as trivial and this one as nontrivial."

That's a very nice daemon! And yet I still wouldn't get on with it. My daemon works completely differently. The beauty of the nontrivial system is that you can get it to run recursively. If a nontrivial system operates in a closed unit, it gets to emergence, eigenbehaviors occur. As soon as you have this kind of recursive involvement, processes become partly predictable, although you don't know why it's happening. That's life. If you take any root, the nth root of any number, and take it recursively, eventually you'll always get one. And now you can invent a nontrivial root finder that constantly jumps between the roots of 11, 22, 9, 1341, and 4; now you set this root spirit loose on the number system—and what is it going to get? One. You know that every time this daemon is active you get one. You want to know how it works? You don't know how it works! And if something claims he can find out, I reply, "How kind, now at last I know something about you!"

However, perhaps what is needed here is time, in some cases a lot of time . . .

Yes, now emergence gets into the game. You let it run, give it time, and out it comes. For example, Karl Müller, he just comes out.

How about the motor of your car, would you also claim that is a nontrivial machine?

Yes, obviously, it's always giving me grief. "What's wrong?" "Oh, I see, the battery." "No, out of gas." "Oh no, the spark plug." "My God, I don't know!" All typical reactions to nontrivial systems. It is our greatest achievement—or our deepest superstition—that we trivialize objects, that we trivialize the world and its "furniture." The triumph of trivialization is humanity's goal. That people constantly cut their fingers, get hurt and fall into despair over it, is thus—I would say—the great problem of civilization.

But in many areas triviality is attractive?

Triviality is marvelous for the user, it's extremely attractive. We live on triviality—and the insurance companies on nontriviality.

The mandate "Trivialize the world" is the implicit consensus at the base of society.[18]

Exactly.

And also partly an epistemological consensus. And this mandate, "Trivialize the world," leads, sometimes to our fortune and sometimes to our misfortune, to actual triviality.

So it is.

And so we ourselves bring triviality into the game, and that's why we have to live with it and pay the price.

One can get voluntarily trivialized. That was all too common under the Nazis.

If we slowly summarize the game of beginnings, then we see we've exhausted two areas. One is that it is not meaningful to get involved in the small-buildings industry and that there are attractive alternatives in the order industry. With regards to the theme of emergence and the limits of explanation, the other important message for the beginning is that we are able to predict and know far less than we normally think.

May I draw another distinction there. Explaining and predicting are two different areas. If you drift into an eigenvalue situation and attain stability, you can of course always make predictions. Some operators converge very quickly, some very slowly—these then are specific questions about the operators—and this problem doesn't seem very pressing to me. Here predictability is important: If I am in the phase where the eigenvalues have calculated themselves, I can make predictions—but I don't know why. That's the problem with "why." Therein lies an unfillable emptiness. You know that something is unexplainable. I can only ever watch what happens when I ask others for an explanation. That sheds some light on those people but not so much on the explanation. Predictability and explainability should be kept distinct. Explainability has nothing to do with predictability; predictions of nontrivial systems are possible, explanations are not.

Normally this is seen differently: One group of theoreticians claim that if I can explain something then I can also predict it, and vice versa. The other group maintains that there are explainable processes that are however not predictable. There is hardly anyone who allows predictability and excludes explainability.

Many say, if I have a model, I can predict as well as explain—except that you have a model that shows that one cannot make predictions if you are dealing with a system whose essential characteristic is nontriviality. That is, a model that deals with operators whose operations change themselves with every operation. Then I can no longer explain, even though I build this operator and know how I have constructed it. But analytically I cannot explain it.

The loveliest example has been provided by good old Ross Ashby, who soldered a little box for his students—we called it the Ashby Box. The Ashby Box was a little box ten by five by four centimeters with two switches, "On" and "Off," and two little lights underneath, also labeled "On" and "Off." Ashby set the box before students who came wanting to study with him: "Here you go, here is this little box, find out how it works. It has just two switches, "On" and "Off," and two little lights, "On" and "Off." Figure it out, please, and then we will see whether we can work together." The students sat there and played with the little box. Because they sat in a room next to my office and because I work late into the night, I was often there at one or two in the morning—and I often thought to myself, "These poor things. They're sitting there next door compiling lists, lists, lists, and more lists." At two o'clock I sometimes went to them and said, "Go home, you can't solve that." "No, no, I'm almost there, I've nearly solved the thing now." "I'd recommend you go get some sleep." "No, no, we can't do that, Ashby told us to figure it out." "But he just wanted to pull your leg and test you." "No, no, I've almost found the solution." The next morning I'd come in at eleven; the man's still sitting there, white as a sheet, hungry, thirsty. Ashby had built these switches so that there were three different logical functions for each switch. One was "and," the other "or," and the third "is." And the internal operations constantly changed themselves.

The students could not have prognosticated this behavior, however.

No, the students could not have prognosticated this apparatus. Someone could have prognosticated it, however, if they had translated the output into the switch position on a symbolic basis, if he had found—or invented—a relationship between the lights and the switches. Whenever they had hit on such a pattern, they could enter it with the switches and after five or ten operations the system would run stably. And the funny thing about it is that sometimes they had the right feeling and thought, "We're already there." But one false step, a step off to the side—and soon it all fell apart again. That's the terrible thing with a machine like the Ashby Box. You think you know it, but even the person who constructed it can't predict its behavior, even though he knows how he has arrived at this step. However, if you can manage to introduce recursions into such a system in a conspiratorial manner or by a hidden path, so that the input/output correlation increases, then the system runs recursively and becomes predictable.

The "drift" in recursion enables prognoses in nontrivial systems. And in a weak sense it also explains why there are particular behaviors, particular limits or eigenvalues.

Naturally, as far as one can describe it. If someone gives me a rule then he is only telling me the rules of representation but not how the machine functions. Ultimately, the machine remains unexplorable. That is not a mythology but textbook knowledge, simply applied to situations that we experience all the time. That is the difference between analytic and synthetic problems. The synthetic problem consists in building a model. The analytic problem lies in peeling out a model as a part of the world and then finding out how it works.

There are different methods to find that out. For example, if you can manage to bring a recursion into this model, to experience it or to live it yourself, then the role I've played has worked. I would say that language is just such an eigenvalue, an eigenbehavior. We speak—and it works perfectly. But there are *many* eigensolutions: Chinese, German, Viennese, and so on. That is caused by the social context that enables these recursions. And recursions are constantly sought after because I can move around in them and feel good in them. Of course there are always disappointments, but these are in the nature of recursions. If I swim, then I swim along in an eigenbehavior, and I think that I have achieved triviality and security. Beneath it, however, lurk the depths of nontriviality.

In family therapy, for example, one sees this relationship in a wonderful way. There's a family that has played itself into an eigenvalue: the husband gets drunk and beats his wife every time he comes home. That has become practically stable; it's predictable. For the therapist, the problem lies in pulling these two people out of this eigenvalue, out of this self-trivialization. If I believe that a family is a trivial system, then I can't heal them. If, however, I know that beneath these apparent trivialities lie deep nontrivialities, I can approach them and try to bring about the emergence of new ways of behaving by setting this ensemble in motion so that a new dynamic equilibrium suddenly arises. And through this one can also see how false and misleading is Prigogine's expression "far from equilibrium."[19] Using my vocabulary, the description could not be more counter to the situation.

Is it not, however, too strong a claim to understand everything, every tranche, every slice of the world, as a nontrivial system?

Obviously not mathematics, obviously not logic—and obviously not those areas that are so constructed that they must remain trivial. Their syntheses are transparent.

But in the consequences for the question of knowledge, the extent of ignorance seems boundless.

I can only refer to Socrates, who said, "I know that I don't know." But many don't even know that. He knows that he knows nothing; that is an initial condition of knowledge; but many do not know that, and that is a condition of second-order ignorance. The game that I'm trying to play is to show that one can know something about one's ignorance, in another way, however, than the reproachful Socrates. I know about which forms I can know nothing on principle. Does that help me? The question is probably different for each of us. It helps me to stay away from wanting to explain something inexplicable, unless I adopt the idea of metaphor, adopt the form of simile, adopt other forms of contextualizing, but not the one of causality. Because then I would have to formulate and use laws that I cannot know. But in metaphor, in parable I need no such thing.

There I say: Just as a camel cannot pass through the eye of a needle, so the rich man cannot enter the kingdom heaven. The parable works because it says "just as." It doesn't say *because* the rich man is nasty, therefore he cannot go to heaven. According to this parable the rich person could naturally decide, "Now I will use my wealth to build a needle that is big enough for a camel to pass through it easily. And then I will go to heaven." There are all these possibilities.

We are slowly coming to the end of our conversation about beginnings. To be a little self-referential, we could modify the motto from Spencer Brown's book: The end of heaven and earth has no name.

By so doing you create a particular understanding of "end." One could of course create other understandings with such sentences; that's what makes them so lovely. I'm delighted that you should propose such a thing. For the moment this sentence is uttered, the end is suddenly orientated, the end takes on an organization. Now I finally "know" how the end is going to end. Or, naturally, one could say that there is no end. To talk about an end, one must know that on the other hand there is no end, or no whatever. Wittgenstein would probably say, "There is no finiteness." One must stand on the other side. The whole idea of the border, of an end then becomes one of poorly chosen language. Or else it simply isn't seen, what one is doing when one talks about the end.

Let's have a go at another paradoxical endpoint at the end of our beginning. We began with, among other things, this proposition: The world contains no information; the world is as it is. One of the points of our last statements was that it is meaningful to get involved in this recursive game because it leads to equilibriums, to eigenvalues. Now one could summarize this in a different form and say: The world contains information. The second part I cannot negate; that would be a contradiction. But one could say it becomes informative to get involved in this recursive game because then the world contains information.

I would play the game like this: I would say that I am capable—through the attempt at closure of an operative unit or the peeling-out of an operative unit through closure—of reaching stable eigenvalues, eigenbehaviors, so that I allow information to emerge for me out of this situation. The information was not in the game to begin with; the results were only won once stabilities were able to develop.

In order for this recursive game to get going, in order for one to get involved, there must be an opponent who also wants to play. From that point of view, one could say, my opponent seeks information because it is meaningful to play with me. That would be a third paradoxical endpoint for our beginning.

I would avoid calling that *information*, because there is something pushing the opponent from within, from which I hope that *my* ability to draw a distinction will be recognized. The distinction is not "there," the distinction is here in *me* or from *me*. And if I come across such and such, then on the basis of my decision I can say, "Now I know that is a mouse and that is an elephant." Then the "information" comes back that I myself have produced. This product usually corresponds with an experience that came about through and in this game. But I would avoid saying that the information lay with my *partner* in this game. Information only comes into being, emerges, grows out of this *game*, in which I feel, recognize or even experience stability.

So the more fitting variation would be: The world contains no information; the world is playable.

Good, the world is a casino. The world is created because I have played it and continue to play it.

That leads us to one of the initial Foerster quotes for the day: At every moment I can decide who I am.

SECOND DAY

Innovation, Life, Order, Thermodynamics

Life cannot be studied *in vitro*, one has to explore it *in vivo*. . . .
Projecting the image of ourselves into things or functions of
things in the outside world is quite a common practice.
—HEINZ VON FOERSTER, *Understanding Understanding*

In every serious philosophical question uncertainty extends to
the very roots of the problem. We must always be prepared to
learn something totally new.
—LUDWIG WITTGENSTEIN, *Remarks on Color*

And God said, Let there be a firmament in the midst of the
waters: and let it diuide the waters from the waters. And God
made the firmament; and diuided the waters, which were
vnder the firmament, from the waters, which were aboue the
firmament: and it was so. And God called the firmament, Heauen:
and the euening and the morning were the second day.
—GENESIS 1:6–8

What will be created on our second day of creation?

*Water: it was created quite literally before everything else. More generally, how-
ever, following the beginning of heaven and earth, today will revolve around*

their continuation. Yesterday we finished with images of an invitingly recursive universe—with the metaphor of the casino. Again and again we returned to variations on the idea that certain pieces—systems—can be peeled out of this universe—and must be. Now, if we examine these pieces or systems, a very important restriction arises. They all operate according to the Second Law of Thermodynamics: In closed systems, order can only decrease and, vice versa, disorder can only increase.

Right away I must remind you of Ludwig von Bertalanffy, who starts out from the idea that thermodynamics represents such an incredible conceptual machine that we ought to use it much more often that we usually do.[1] If I have a heavy hammer or machines, then I can build things that I could not build without the hammer or without the machines. After his initial dismay that thermodynamics were only used by engineers, by builders of steam engines and the like, Bertalanffy's fascination led him to the idea of using thermodynamics for the analysis of living organisms.

At first there was a problem with the first requirement that is usually necessary for the application of the Second Law and that cannot be used in the study of living beings. Because I have to presuppose an adiabatic—impermeable—covering, outfitting for the system I am observing to ensure that it remains an energetically closed system. If I do that with a cat and put an adiabatic cover over the cat, then, unfortunately, five minutes later only the cover will be intact—and there will be no more cat. Thus I lose the essence of "cat-ness" if I wrap it in an adiabatic compress or cover.

Therefore, I must open the system that I want to analyze from a biological point of view; I must drill a hole in this cover so that energy can get in. Bertalanffy was the first to draw our attention to this: "If I want to apply thermodynamics to living organisms, then I must have an energetically open system." If I have an energetically open system, however, then what can thermodynamics, what can classic thermodynamics do for me? And he quite rightly concluded, "A lot." When the formula claims that the change of energy going through system equals exactly zero, that only indicates a simplified case, only represents a special instance. I can also write that dE/dt—the change of energy over time—equals 25 calories per second. What happens then? What do these different equations look like now? Here the energy of the system is not a constant, the change of energy in the system does not equal zero. Now, how can I use the fundamental equations of thermodynamics for cases in which the change of energy has a certain positive or negative magnitude?

And so Bertalanffy began to write these new equations, and through these began to found a theoretical biology. This step was crucial for me as a young man. "Aha, Bertalanffy understands this, wonderful, let's build a mathematics that deals with the thermodynamics of open systems." In what ways does this new view of thermodynamics fit the classical idea? Well, Boltzmann's classic idea was to say, "I don't just want to busy myself with simple thermodynamics and only ever produce equations for energetically closed systems." Since, after all, one can write a tremendous number of such equations. And in many cases I can also claim that a big steam boiler, which is wonderfully insulated so that the heat does not pour out, may be treated approximately as an energetically closed system. That is done all the time in physics. Boltzmann then applied this peculiar idea of thermodynamics to the world, to the universe. Although we see no adiabatic covers anywhere, the world is still probably a finite system. If it is a finite system, then we can also claim that it has finite energy; there is nowhere for outside energy to come in, and there is nowhere outside the world where energy could disappear to—the energy in the universe is constant.

As soon as I get to that point, I can apply the Second Law of Thermodynamics with complete success because I am dealing with an energetically closed system. And this way one can use it to make all kinds of predictions—one can therefore use thermodynamics to say that order, that relations between elements in the system, cannot exist in such a way that elements drop out or disappear, but that the order I observe will increasingly slip away from me. How did Boltzmann see order? He very rightly regarded order as the difference between two objects. As soon as I say, "That is an ashtray," and, "That is a tape recorder," I have given the world a certain order: the ashtray is an ashtray—and likewise the tape recorder is a tape recorder.

But if, while we're sitting here, the ashtray suddenly starts to look more and more like a tape recorder and the tape recorder more and more like an ashtray till in the end all I have before me is a grey, black, silver mass, then I would say, "Oh, no, all order has been lost." What does this concept of order, which I have poetically clarified with ashtrays and tape recorders, look like in physics? I take two containers, one is A, the other B—A is hot, and B is cold. If I put my hands in them I can easily tell the difference between them: "Ouch, that's the hot one," "Brrr, that's the cold one." So far, so easy. Now I take these two containers and I connect them so that an exchange of heat can take place between them. I wait and wait, and after a time I can no longer distinguish the hot from the cold; they are both the same temperature. The fundamental prerequisite for this equalization is that the wall between these containers be opened so that the molecules can move back and forth. That is called "diffusion." And in this diffusion the movements level out.

Boltzmann was the first to think about this process intensively, and in so doing, incidentally, he advanced the idea of molecules. As you know, this very concept was severely contested by Ernst Mach and many other contemporaries. One time, Mach was sitting in the hall where Boltzmann was lecturing and as soon as the word "molecule" left Boltzmann's mouth, Mach shouted, "Show me one!" With this game of molecules, of course, he could demonstrate beautifully how temperature equalizes through diffusion. But what happens with the observer here? Now, this point is important. The observer explains, "Order is lost." The observer is perturbed. There's a very nice play on words: "Diffused is confused." The observer no longer knows which is A and which is B. Very well, Boltzmann took this process of the two physical containers, in which the molecules from container A flow to container B until eventually one can no longer tell the difference, and applied it to the universe. "Here is a beautiful star, there is a fog in which atoms chase around after one another in a confusion. If no energy comes in from outside apart from the suns that shine, radiating their energy onto the planets and the earth, then after a time an equalization will occur: the sun will cool and the surroundings warm up. After a time I will no longer be able to distinguish one from the other. In this way, the whole universe will die a so-called heat death." Why is it called heat death? Because the energy difference that exists between warm and cold bodies has disappeared.

We use this energy difference to drive a train from A to B, to build something or to create something. 100C on the inside, 20C on the outside: industry lives on that. If this is no longer the case, industry ceases, we all lose our jobs, everyone dies, nothing is there any more. That was Boltzmann's idea of heat death. Then Bertalanffy said, "Wonderful, my dear friend Ludwig, I won't look at systems that are thermodynamically closed, I will look at systems that constantly consume energy and give it off again, which radiate heat or digest it or whatever you want." And that was already happening in 1923 or 1925. I'll explain it now in a simple form: This basic idea was already there when Prigogine, the great Prigogine, spoke of "dissipative systems" or "dissipative structures" forty years later, in 1960 or 1970.[2] When I stumbled across his "dissipative systems" I could only wonder, "Why does Prigogine get a Nobel Prize for that?" If you only look at the output, if you only look at the excrement that a system produces, then it is a dissipative system. But I am interested in the caviar or the champagne that this system is eating and drinking, thus, in a thermodynamically open system from Bertalanffy and not a dissipative system in the sense of Prigogine. How— now, regrettably I must use the world "shit"—can a system shit if it doesn't eat? And yet for "dissipative systems" one gets the Nobel Prize, but Bertalanffy did not get a Nobel Prize for his "open systems." That's Nobel life.

Bertalanffy has certainly laid the foundations. From the beginning he was using equations in which entropy and thus the change of entropy over time (dS/dt) were not always dimensioned so that they grew but so that they could also diminish. Why? Because energy is pumped through the system. And now there are various aspects that one can use to revive the emergence of distinctions and differences. Numerous possibilities stand open: Prigogine and his Belgian colleagues conducted some very amusing experiments. If, for example, you let liquid draw, then you suddenly see patterns developing, turbulences, and so on—because it is a thermodynamically open system. Now I'll be a scientific politician again: If I just call this kind of system "dissipative," then I don't realize that it receives tremendous inflow because energy is being pumped into it. Yet if I describe this system as thermodynamically open in Bertalanffy's sense, then I get the feeling that I am giving the problem the proper, full perspective: I'm looking at both sides of the process, not only at what comes out but also at what goes in.

Some time ago, I often used to think with friends, one of whom was Gordon Pask, about the idea of self-organization. And of course self-organization comes into play at once when one considers the Bertalanffy interpretation of thermodynamically open systems. Because then one asks, "What can energy do if it's pumped through this system?" And there the most diverse aspects can be demonstrated. First I'll give you the "Foerster aspect," which in my opinion is the simplest. The Foerster aspect claims the following: "If in a system there exist building blocks or elements that can enter into certain relations, thus showing a potential structure of relations, then they need energy in order for these structures to be realized." If I have some components—for example, balls and bowls—and I pump in energy by shaking them, then the balls will go into the bowls. Or: If I have little hooks and I shake this system, then I'll get a chain. And if for example you work in a garden, you will also, to your great annoyance, come across self-organization; there's nothing you can do about it. You constantly see self-organization: You walk with a dog on a leash and soon you'll get caught on a root or a branch.

I would, by the way, make the following recommendation to nuclear physicists: Don't think about little balls that won't have anything to do with one another; think about little hooks, the whole universe is made of little hooks. And if you shake them, then the hooks join together, especially if they're those fishhooks that are bent at both ends. You've got to give it a little shake. Everyone can try this out in his or her room. Put fishhooks in a pot and shake it. Then take them out. They're all connected. Marvelous self-organization. All I need to do is shake. And nothing else. That means that if I let energy flow through a thermodynamically open system then the structures, potential structures that exist, will be realized. And in many different ways. I cannot predict how this hook system will look in the end. I can only say: I can pull it out at the end. But how A will be connected with C or M, I cannot know; that means that I can't tell whether an elephant, a frog or a goat will come out.

If we clothe it all in metaphors, then first there is the metaphor of heat death. There an image of a final stable state of complete disorder comes to mind.

Yes, that is the question. The word "disorder" is already the question. What does "disorder" actually signify? But please, let's continue with the metaphors.

A counterimage, a picture of an ever-increasing order, would be a counterexample in which ever more complex forms emerge. If we connect the two, then a comprehensive picture emerges. That gives us a picture like the Escher print Order and Chaos, *where one sees, in a jumble of garbage, something very beautiful and sophisticated.[3] Here islands of great order coexist with surroundings of even greater disorder. Is that a metaphor you can live with?*

That's the beauty of metaphors, you can always make something of them. I would say that is the Karl Müller metaphor for working his way handily out of the order/disorder problem. May I ask a question at this point: Who sees this order? Who says, "That is or is not ordered"? There must be someone who makes those classifications. Now we have to get to the observer who says, "That is ordered" or "That is not ordered." At this point he says enthusiastically, "Look at this beautiful thing." He has recognized some order; otherwise he would not have said "beautiful." "The chaos there is quite terrible"—he sees, or better, he projects what he sees onto something that he can deem to be a disorganization-structure. Another observer doesn't get that; he does not "see" these orders and disorders, for the time being he does not have the metaphorical wit, the power, to see these structures of relations. That's why I'm so proud that from the earliest days I proposed to measure organization by putting an observer there and instructing him: "Measure the structures of relations that you can recognize and order." In information theory there is a lovely expression that is called redundancy: If right from the start I describe one and the same object as so and so and so, then every further description becomes superfluous.

Redundancy represents a measure of order, and information theory has developed measured values for redundancy. For this one must count the structures of relations. If their numbers increase then redundancy becomes greater, then order increases more and more. Order is perfect if I can use one proposition, one point, and through that all the rest of the structure is explained. With that I am also able to make perfect forecasts. If I say, "That is a cubic structure," then with the length of one side

I know everything about the structure. There the order is very great; with one proposition I can immediately draw conclusions about other propositions. I introduced this order principle, and with it we can write equations that tell us whether a system is self-organizing or not. Then if the organizational structure or redundancy increases, then more and more order is also created. And I cannot increase order through anything other than a supply of energy.

In this view, then, the crystal becomes a highly self-organizational system?

It behaves as a highly self-organized system. The question is just—how did it come about? That requires some explanation: The crystal first existed as swimming particles of salt. Then through shaking I introduced heat, which is also energy. Through the addition of heat, crystallization occurred, salt crystals formed. If on the other hand I leave the crystals in this soup to crystallize themselves, then we have an endo-energetic system in which it gets colder and colder through organization until it comes to an end. If I add a little energy to this system, however, then crystallization continues. Then one merely has to pump energy into it and organization increases, as long as the elements that compose this organization are capable of forming such structures of relations.

Two important questions. The first concerns the durability or "sustainability" of self-organizing systems. There is for example a very famous sequence in the Marx Brothers film Go West *in which they are going back east by train and in order to keep the train alive, that is, to keep it going, they begin to break the cars apart to get fuel for the engine.[4] The train goes very fast, but there's a cascade of film cuts in which the cars are demolished bit by bit—and in the end all the passengers are standing outside and all the fuel is used up. That means there's a connection between self-organization and self-destruction?*

Once a system is an open system but there's no more energy left to be pumped through, the self-organization ceases. If you take energy from the system itself because none comes from the outside, as in this typical case of the self-destruction of a self-organizing system, then this necessarily means the end of this system. If however energy is used to destroy another organization, then this assemblage becomes a "self-disorganizing" system. If you read my early paper, it describes exactly this point.[5]

The second point is about measures of order. That what you have proposed with reference to redundancy is only one possible measure of order. Essentially there is a multiplicity of such measures of order.

What do you want to know—shall I enumerate some others? The other best measure of order also comes back to a fundamental point: Order is a problem of description. If we claim that this is better organized than that, then the descriptions must "show" this. In the initial discussions of problems of self-organization, descriptions weren't touched on at all. Lars Löfgren, a member of the Biological Computer Laboratory (BCL),[6] had an ingenious idea: He set the step from observation to description, a crucial step in my opinion.[7] He said, "If I have a system and I describe it in state one, in state two, in state three, and so forth, and it relates to a system that has the tendency to or rather the possibility of changing itself, how do I change my descriptions so that I can claim, 'Here it has self-organized,' 'There the system self-disorganized'?" And because he is a good mathematician and a Turing specialist who knows how to handle Turing machines, Lars Löfgren found an excellent method for evaluating the change in description of a system. I would like to briefly explain what Löfgren's idea of description looks like.

A Turing machine consists of a strip of any length. On this strip there are drawn little squares that are as large as the strip is wide. In each of these squares one can write a symbol. Here, to keep it simple, a symbol is either 0 or 1. One could of course omit the 0 and have an empty square, 1, a point, a star, a symbol of an elephant or whatever. But Lars took 0 and 1, the usual digital arrangement. For every description that one draws up, one can find a code that can be written on this strip. One can develop a binary numerical code with 0 and 1: The letter *T*, for example, is defined as 011000. Once I've done that for the whole alphabet, then I could code "jellyfish," and then jellyfish is also on the strip. If I want to read that again, then I give the machine the order, "Read that," and the machine responds as follows: It reads the first square, looks to see whether it finds 0 or 1, notes a 1, then the strip is moved to the next place, is read again, and so on. Furthermore, the squares also control the behavior of the machine: Should the strip be moved forward or backward, should one of the signs in the squares be deleted or printed, and the like? Turing has shown that as long as you have such a machine at your disposal—that is, a Turing machine—you can describe everything that you want to describe.[8]

There you see the parallels between the Turing machine and the McCulloch-Pitts theorem.[9] The McCulloch-Pitts theorem claims, "Everything that I can describe can be represented in a neuronal net." Eventually it was John von Neumann who saw that Turing machines and McCulloch-Pitts machines are identical systems represented in different languages. He was the first to show the equivalence between McCulloch-Pitts and Turing—a tremendously important result.[10] Well, now I come to making a description of a system with the help of a Turing strip, and I will sort it out. Now I can determine whether there are redundancies in this description, meaning: Is there something in this description that I can omit because it is redundant? I can just keep eliminating more and more whenever I encounter these redundancies—till I get the so-called shortest description.

Now I give the machine some time and energy—maybe I give it a little jiggle, deedle-deedle—then I begin once again to calculate the shortest description. If the earlier shortest description is longer than the later shortest description, then the system has organized itself. Why? The shorter the description produced, the more redundancy has been found in the system and the higher its organization must be. Lars Löfgren ascertained that in an excellent paper, "Recognition of Order and Evolutionary Systems."[11] I recommend that everybody read that. I use these references to Lars Löfgren again and again. Why? Because in a very precise way he introduced the observer as describer into the whole argument about order, unorder, disorder. In my opinion this is a decisive step toward an understanding of the order and disorder problematic. If it is not understood, then the discussions about it become disorganized—and lead to unenjoyable nonsense. People don't recognize that it's all a problem of description.

There is also, however, a Foersterian connection between observation and description. "The information in a description depends on the observer's ability to draw conclusions from this description."[12] How is this ability, which must lie with the observer, anchored in Löfgren's idea?

That fits the Löfgren model exactly. I must find a code, and with this I will equip our jellyfish with a dash-dot-dot-dash. The language with which I do this can be freely chosen, which is, by the way, a second Löfgren idea that fits in with our discussion right now. The length of the description of an observation depends on the language with which I want to describe the observation. The language that I use has an inbuilt redundancy, which is then reflected in the shortest description. What therefore does the use of language mean? The use of language means that there exists a semantic structure of relations in language. This semantic structure of relations also finds expression in the conclusions I can draw from a description.

Information in this sense is double-sided: It relates to a single description but also to the resource-language as a whole.

Exactly. Those were also my short remarks on self-organization. And in my opinion the decisive step—the Löfgren step—consists of shifting the problem to description and language. The fun lies partly in the consequences of this step. People quarrel about the value of the order, blahblah: "Tell me what language you're using? This or that? Fortran? Or another computer language?" It immediately changes the situation. In my little paper "Disorder/Order—Discovery or Invention," I have tried to describe this issue as clearly as possible.[13] In it I incorporated Löfgren as the key to this whole problematic.

Let's assume further measures of order, which could easily be constructed. It's possible to imagine contradictory measures of order by taking the position that order increases with the length of the descriptions because systems that take longer to describe are apparently more complicated, and order is associated with complexity.

I'm afraid you're confusing two problems. At first I spoke about a certain configuration that can be described in various ways. How this description comes out is a language problem. And if I find that a description can be shortened because certain redundancies are apparent, then I concentrate on the shortest possible description of the system. Then comes the second point. How many steps do I need to describe this description— that concerns the length of the description on the strip. There one could say that length has something to do with complexity—if we don't yet have the shortest possible description. Could I identify the length of the description with the complexity of what is being described? Nothing seems to speak against it: The length of the strip of the shortest description required by the system gives me a second measure, one which I can use to compare two systems by taking the shortest descriptions from both descriptive systems. If one was short and the other long, I can say one is complex, the other simpler. Would that be a satisfying solution? I'll introduce a new word; I won't use complexity for order. I will use two expressions: one is order, the other complexity.

I totally agree with that.

Let's move on. What's your point?

I am simply introducing my apparently disorganized person whose definition of order is precisely the opposite of yours and Löfgren's. You see order as changes in the shortest possible description. An increase of order or self-organization has occurred if this description becomes shorter. My disorganized person sees order as a phenomenon that has increased when the descriptions get longer.

Very daring, Mr. Müller—or Mr. Jones, in this case. Mr. Jones. Very nice, if that's how you want it. We live in a free country. And in a free country freedom of speech and thought reign. Why not? I don't want to argue against that at all. If Mr. Jones finds that he likes to describe order with more words than me, then my telegrams are cheaper.

Moreover, my disorganizer describes as self-organizing systems precisely those areas that you see as not self-organizing, and vice versa.

That's his problem, he can go see a psychiatrist as well, there's nothing stopping him. I'm not going to say, "That's wrong!" If he has fun and feels good about it, then he should be happy after his fashion. I see it like this: If something is organized, it is therefore easier to move from one point in a system to another because I know that this point is there. If I know of this pen that—if I open it—I can write with it, then that's an advantage over a situation in which I don't know that. If an object is organized so that it's packed with redundancy, it can have advantages. I know that if I pull here it will open. I know that if I write the letters will be black. All these things are because of a certain order that exists or, I should say, is known. This pen is ordered in the sense that I already know how it writes. Of course, that depends—this is Löfgren's step—on my having a general knowledge of how one can describe a pen. If I could only describe it as a little rod, a black little rod, a straight little rod, and so forth, then I haven't understood the possible uses of a pen at all.

Another malicious argument would of course be that the systems that are simplest to describe are trivial machines.

Sure.

Why should it be precisely those that achieve the highest measure of self-organization?

Why the word "self"? There are two problems. One is called self-organization; the other is called organization. Organization lies in my being able to look at a thing and say, "I can go from there to there because I know how I'm going; it fits the descriptions and the redundancy." The second problem consists of how the redundancy got there. What was the starting point for the uniformity that I now see in this organization? And then the question is, "Was there someone else at work? Has that someone hammered, nailed, and screwed this organization in with a screwdriver, hammer, and so on?" Then someone else would have organized it this way. Now I can pose the second question: "Those people with hammers and screwdrivers, were they also organized from the outside?" No, they had their buttered bread and their beer and banged and hammered and nailed this machine together. "Aha, these people therefore make up a self-organizing system. There was no one from the outside telling the men what to do; the only thing added was energy, namely, buttered bread and beer and other things." Only now does the world "self" come in—self-organizing and self-organization.

One represents a process, the other a description of a situation. So: process and situation, that's what I distinguish between. Now, what can a process refer to? As soon as you say "self," you're dealing with a process; as soon as you say "organization," you're talking about a condition, or rather, the description of a condition. If you're thinking, "If I see such and such, then it's a self-organization," you actually cannot say that yet. You don't know from where this system has emerged. Perhaps it's a self-disorganizing system, so that unfortunately it only looks how it looks. I don't know if I've made myself clear—I'm talking only about vocabulary, nothing but vocabulary.

How can one keep track of the differences between organization and self-organization?

I'm talking about two different problems. In one I'm talking about a condition, about something that is there. In the other I'm talking about a becoming, an emergence, a process. With a process, the word "time" comes in as well. In the other case the word "time" doesn't come into it at all. As soon as I have the word "self-organizing," I'm talking about how the process developed so that this organization emerged, regardless of whether it was produced from the outside or whether it was created through one—and only one—jolt of energy.

Perhaps we have to introduce the terms "auto-organization" and "hetero-organization." In the case of hetero-organization, some mechanisms from the outside increase the redundancy of the system: If you screw in a screw, then these parts won't fall apart anymore; if you hammer in a nail, it will hold together, and so on. Is this nail, is this screw brought into the system by someone outside the system? Or were the nail and the screw already there in the system? Did someone shake this system—so the screws screwed themselves in and the nails bored into the wood? In this case I would speak of "self" and also of auto-organization.

Seeing as we have been busying ourselves since yesterday with slices or systems that have been cut out of the "world," can one say of them as well that they are auto-organized or hetero-organized?

Sure.

And are these ascriptions and border drawings made by the observer?

No one else could do it, one has to look at it, one has to look at it.

But among these observers a legitimate diversity can exist.

Sure, definitely.

Thus these classes can be built in a totally contradictory manner, as shown by the example of the self-disorganizer. The normal thing would be for these different measures of order and perspectives to coexist?

If these observers don't intend to speak to each other, then they could each adhere to their respective measures of order. In time, they might even find a common language—if there is time enough under the sun—and enough sausage, spinach, and beer to keep them fed.

Let's move on to the second big subject area. If one engages with a self-organizing system, then there are—as you pointed out—two principles for how self-organization comes about: "disturbance from outside" and "self-organization from within." Could you say something about both of these principles of order?

I've made these remarks vis-à-vis other conclusions. They are therefore not out of context. The context comes from a splendid book by Erwin Schrödinger, *What Is Life?*[14] That book impresses me greatly because in it he lays out principles of order using clear steps, in which he distinguishes between statistical and natural-law structures. Among other things, he has introduced the idea that life and its dear complexity emerges because order is imported from the environment into the system. The metaphor that Schrödinger uses is the "food-order" of the universe—and the answering question is, "How does one eat up order from the environment—and what kind of stomach does one need to digest order?"

This metaphor is outstanding, quite enlightening. You eat some spinach, for example, which already contains all sorts of things, you use them here or here or here to build molecules—and already you have made your life much easier. What I stress in my definition of self-organization or organization is that it lies in the extant potential structures of relations. You have, as in the example of the hooks, a structure of relations in which hooks connect to form a chain, that's the simplest case. Now I'm going to connect the word "potential" to my structure—the physical expression "potential" refers to a possible action of to a possible distribution of energy. I'll bring this concept of potential into the inner structure, where I can speak of "potential structures of relations"—two hooks, three hooks, or much more complicated chains. Now I no longer need to import order from the outside; in the given case, my system does not have to take in more order from the outside.

Schrödinger's image is a lovely metaphor—"Order from order"—order emerges from order. I, however, would say that it does not have to be that way. Order can emerge from disorder. And in engineering the expression for disorder is *Rauschen* in German and "noise" in English. I hope that it's possible to see that I can also create order from disordered noise—noise also leads to energy. This principle is called "order from noise." I've explained it further in an article, and this idea was seized on with great delight by many people.[15] In the case of "order from noise," I don't have to import order so long as potentials and potential structures of relations exist in a system through which energy is flowing.

What would be some concrete examples for the metaphor of hooks—and their chain-formation?

Everything, everything, everything—we're constantly finding these hooks. There's a bowl in front of you; it's a "preorganized" object with a specific structure. You throw cigarettes in there—and they stay mashed up together there. You could, of course, throw them somewhere else instead. But there's this "key structure" here, which allows you to create a redundancy so that these cigarettes lie very close together. You could, of course, have thrown them all over the place so that no one would know how these butts were distributed. Through this ashtray a potential organization exists that you are more or less happy to use so that disorder is not increased but rather decreased.

These hooks are found everywhere, everywhere, everywhere. This book here has a bookmark, which you use like a little hook because the pages prove "potentially turnable" and you can shove a little hook into them. Thus a "prestructure" exists that you can use to find the place again, which you would not have found so easily without your bookmark, without your prestructure. In the case of the bookmark and the book, the redundancy lies in your being able to dramatically reduce your search operation in this way. When you tag a page with a bookmark, with a little hook, you know you'll "automatically" get the last page you read. There's no need to leaf through the whole book, one page after another, because there's this bookmark stuck in there: I know where I have to look.

Is there a difference between self-organizing systems and living systems?

Living systems represent a beautiful instance of self-organizing systems.

And is it possible to find self-organizing systems that are not living systems?

As soon as some realizable structures exist, the system organizes itself, as long as there is no external "great organizer" keeping this structuring going. In the case of the organization of the solar system, what was and is the realizable structure? The structuring element is called gravitation. Johannes Kepler noticed a redundancy in planetary movements, which "forced" the planets to move elliptically around the central body, and so it's an organization. And then one Isaac Newton comes along and says, "Yes, I can explain that!" And he invents a cosmic nuisance that pulls these planets back and forth. "By God, the apple! That should also be applicable to the planets!" And so now we have an "explanatory principle," gravitation.

So far we have discussed the question of order and disorder in some detail, as well as the problem of measure in several variations. Several times the concept of complexity arose. The conceptual pair "complexity and simplicity" is not, however, identical with the conceptual pair "order and disorder?"

What's this identical? That is simply a question of language and nomenclature. I wouldn't call a wire a piece of wood—unless maybe people were saying, "That's a wiry piece of wood." In this case I would reply, "Let's forget the difference, call it even!" I find it convenient to maintain a distinction between complexity and order.

We have a unified dimension with the dual poles of order/disorder. Now how does the dimension of complexity and simplicity relate to the order/disorder dichotomy?

First of all, I wouldn't consider your order/disorder dichotomy to be a dichotomy.

The shortest possible description and the longest possible description—that is one dimension.

Good, that's in the same area. And now someone comes up to me, for example, and says—

Complexity.

Edgar Morin comes up to me and gushes, "*Complexité, complexité!*" To which I respond, "Calm down, calm down, let's have a coffee!"

Living systems represent a specific class of self-organizing systems, which should stand at the center of today's conversation. If asked to describe the question "What are living systems?" as briefly as possible, what would you say?

I don't know whether one can express that in a shortest possible way. In my article "Notes on an Epistemology of Living Things,"[16] I tried to write the story so that it contains itself, that is, to describe the question of life not as "*There* is life," but rather to allow this question itself to become lively. The life-answer shouldn't be, "That dog *there* is alive, that stone *there* is not alive." The answer should rather make it possible to become immersed in a world of description, a world of language that brings life to life.

I can't say whether my article was really successful, but at the time I was happy that this circular, almost Wittgensteinian structure occurred to me. It's similar to how in *Tractatus* there are propositions supported by subordinate propositions, and the beginning of the article implies the end, and the end the beginning. That, by the way, is an essential difference to Wittgenstein, because he begins here and ends there. My intention, on the other hand, was not to write so that it begins here and ends there, but rather so that the end here leads back to the beginning here—and the whole thing stays as a circularity of here.

In essence I always hoped: If I talk about life, the life should become something that is describing, something that can describe itself. Whenever an interactive describing forms an organizing system with something else and we experience a constant interplay, then a constant, potentially inspiring dynamic develops. With it I might manage to stimulate life in the reader. My hope was to describe a situation in such a way that the description itself draws the reader or the listener with it into the dance of describing, of listening, of reading or of creating. All good poets can draw me into their world and entangle me in it. I admire artists because they can tell me what it's all about when we talk about life.

Tellingly, one of our mottos for today comes from the fount of this epistemology article: "Life does not allow itself to be studied in vitro, one must feel it out in vivo."[17] This proposition shows the concept of life to be a typical second-order concept.

Exactly. In our cooperation as well, if we could, the essential thing would be to give reference to the so-called second order. The moment I refer to second order problems, I also point out that the world, the description-world, changes as soon as I draw attention to the fact that we describe the world that we describe. All at once I can ask, "How does a concept function? What is the 'concept' of a concept?" Fissures immediately emerge—for me as listener and as speaker—that don't come up otherwise when I speak about concepts. If I apply what's been said to life— "Where is life 'here'? How does life live itself?"—then of course for a moment it sounds like an excuse. But if you think it over a bit, then it's not an excuse but an invitation. I have tried to express that in the opposition of *in vivo* and *in vitro*. It makes a difference, whether you're sitting in front of a test tube or whether you yourself are dancing-with and playing-with. Thus if someone asks me about life, I answer shortly, "Let's just live! For once, let's just live life now."

That, however, also implies a substantial aspect shifting for any biography?

If, for example, I was invited to play along with my autobiography, then for me it would always be a bioautography, that is, a self-description in which life itself becomes alive—thus, a bioautography.

There is a classic description of living systems as a specific type of self-organizing systems, namely the autopoietic idea: The decisive characteristic is that the interactions of a system themselves coproduce the component parts of the interaction.

The concept of autopoiesis, which Maturana and Varela developed in cooperation, in interplay with the strict BCL-logicians like Löfgren, McCulloch, created a configuration that referred to such second-order phenomena. Tellingly, the autopoiesis article originally appeared as a BCL report and was, incidentally, stolen by a publisher in Boston without any BCL references.[18]

When I myself was in Chile, we were thinking, "Could we write a living computer program?" If we could write such a program, it would become clearer what we're talking about when we talk about a living system. I find the autopoietic form of self-organization, in which the elements constantly reconstruct or construct themselves, very important. There we were in Chile, sitting together, arguing, sitting together, arguing, sitting together, arguing. . . . When the first manuscript was finished, I had to leave Chile immediately because Pinochet had just had his putsch and Allende was assassinated. Mai and I left Chile a few hours after Allende was assassinated. First we went to Mexico, and I had the autopoiesis manuscript with me. One of my students, Ken Wilson, and I translated the Spanish manuscript and published it in *Biosystems*—that was the first English publication.[19]

Of course I found the basic autopoietic idea, Maturana's basic idea, very important—I've always strongly supported this conception and even today consider it to be a very significant definition. But Maturana was centrally concerned with what would then be a living system there. And I like his definition of what a living system there is very much. For myself, however, I try to construct a description so that I involve the reader, so that together we form a living system in which we carry out the game of writing, reading, talking, understanding, answering.

How then—if one used this distinction—would living machines, ourselves included, distinguish themselves from other, nonliving machines? Can one find criteria "here" and "there" for whether, for example, this typewriter is or is not a living object?

Clearly you are asking the there-question: "Is 'there' a living type-writer." Then I would suggest Maturana's autopoiesis definition. If I claimed that the typewriter there was alive, then I'd apply the criteria for life.[20] I would calmly put them forward and ask you: Use the Maturana tests on this typewriter. And if the tests turn out positive, then you have before you a living typewriter, otherwise not! That is my suggestion when we ask there-questions. But I see the problematic from a somewhat skewed perspective—I see the problematic of form. I would, for example, like to ask and ask back in a dynamic form. You ask me, "What is life? Is that a living typewriter?" I answer, "Dear Karl, in which form would you like to see my answer so that you might go home satisfied?" There is the dictionary form—we flip through the dictionary, there is an entry on what life is—and with that I have given you "life"; or the poetic form—we find ourselves a poet who glorifies life. You ask "What is life?" and I answer, "Take the form of the poet who writes a hymn to life!"

I am now interested in the specific form in which we want to cast the problem "What is life?" so that you can go home satisfied. If I gave you the little book *What Is Life?* by Erwin Schrödinger and you read it through to the end, you would say, "Now I know what life is." Or you would say, "No, that isn't really what I wanted. What I actually wanted to ask was . . . etc., etc." I'm trying, with these remarks, to draw attention to the fact that questions and answers exist in certain forms, in those that satisfy the questioner and the answerer, and in those that do not satisfy the questioner and the answerer.

I think that the question "What is life?" can offer no forms that will satisfy us, or at least none that will satisfy everyone. Because the question is already once again a question about the question. Now is this satisfying for the questioner or not? Or, with regard to this question, is the unsatisfactoriness such that we become more interested in the question of life? In this case we've won the game—because then life stays open and will constantly be led back to life. Am I making myself at all clear?

You mean that for that kind of there-question there are no satisfactory here-answers?

If I don't know what the here-answer is, then it refers to the form of a question. I don't know what the here-answer could be. Thus I ask, "How would you like to see the here-answer?"—And you can't tell me that; otherwise you would not have asked. If someone asks me, "Is there life there?" then I call on Humberto Maturana: "Please repeat for us your life-criteria; then we can test whether or not this typewriter is alive!" It comes out as yes or no. For me, however, the problem presents itself differently. I just want to stress that. For me the problem lies in which form you would like to see my answer. That is identical to the question, "Are there here-answers for there-questions?" My reaction to that: "If you can tell me what a satisfactory here-answer looks like, then we will have successfully created such an answer." But I claim that you cannot tell me how this here-answer should look. If you now say, "But I know the here-answer," then I'll be very grateful—because now we can talk about the weather.

Let's try to push these here- and there-questions even further, namely in the context of your article on molecular ethology.[21] For the self-organization of living systems—and also for our conversation—it is vitally important that something new should emerge from time to time. What makes up the repertoire of operations that is necessary for this? One operator formulated in the classic evolutionary literature was random change, mutation. That, however, would only be

a very weak operator. At the same time we know of a whole series of other opera-
tors, for example, the merging of symbiotic constellations.

Lynn Margulis has dealt very extensively with these problems of symbiosis and of "merging."[22] I've always liked the Margulis idea a lot. It ap-
pealed to me from the start simply because all of our orthodox colleagues
wouldn't listen to her and didn't want to engage with her. Here, however, we are touching on a scientific-political question. First, though, to
the random variations. Chance always crops up when I don't have any
explanations handy. The great French biologist Jacques Monod placed
"chance" and "necessity" in opposition.[23] This form of comparison doesn't
seem right to me. More correct or more useful would be the contrast of
necessity and freedom, that's my favorite for comparison, not chance. It's
chance only if I don't know why one thing has come to another. I go into
the street; suddenly, along comes my friend Karl Müller. "That must be
chance!" If you look at your ticket and my ticket, then it becomes clear.
"We had to meet each other at 17:23 at the train station." Thus the suitable comparison is not "chance and necessity" but rather "freedom and
necessity." In my poetics of life it is essential that freedom should exist.
Again and again, every moment, this self-organizing system has the
choice to function like this or this or this. There is always the possibility
of moving here, going there, turning the opposite direction, and so on. If,
however, necessity rules the whole system, then we just have a mechanical
typewriter, then we have an automaton before us.

   If I use the Pask language, then there are several types of composition.[24] In logic, one speaks of composition when two or more things
come together to form a new unit. We distinguish between superadditive and subadditive compositions, as well as compositions in which
neither addition nor subtraction takes place, and instead the objects remain isolated alongside each other. Thus the idea, as an evolutionary
principle, of getting a superadditive composition, such as a coalition, is
very important in my opinion. It is very important because through cooperation or symbiosis new possibilities will arise that the original elements did not possess; these possibilities are only available if they function
together. The repertoire, as you called it before, the repertoire of operators doesn't suddenly double or quadruple, rather it takes on qualitatively different dimensions.

Perhaps we can summarize this point like this: A very important characteristic of living systems—and also of our conversation—is that they possess a multiplicity of operations for creating something new.

I am stubborn and I am going to emphasize the form-question yet again. "With which form will my dear Karl be satisfied?" In my game I would not be satisfied with the form of enumerated operators. I prefer the Albertean formulation: "What is the here-answer to the there-question 'What is life?'" We possess no here-answers. And with that I'm trying to keep the question of what life is alive.

In my summary I actually meant something more trivial: Living systems—from unicellular organisms to plants, animals and people, as well as interactions and conversations—are characterized by, amongst other things, their many possibilities for creating something new. Take for example inversion, which can be observed in the field of genetics as well as in conversations. In the 1960s, things were "hot," whereas today they're "cool" or "super cool."

Beautiful, beautiful. I would just say, "Inversion always just comes down to the same thing, nothing new comes out of it; it's always just the same." May I bring up Wittgenstein again? There is a proposition in *Tractatus* that states that a proposition *P* and its negation are talking about the same thing: *P*.[25] And I always urge my friends, the revolutionaries, to take this point to heart; I say, "If you're shouting, 'Down with the king,' then you look like you're on the king's payroll because you keep bringing him up. The main thing for a king is being mentioned, staying in the game, and it doesn't matter how it's done. If you really want to get rid of the king, then you shouldn't talk about him anymore; only then will he disappear."

This point has important consequences. In this context I'd like to remind you of a significant work by Gotthard Günther that is rarely understood correctly.[26] And why is Gotthard Günther so little understood? Common logic knows the truth-values "true" and "false," which can be interpreted as 0 or 1 and can be extended to a value arithmetic in which the particular symbols are input values for functions. Gotthard Günther spoke, however, of an additional value, and readers or listeners acquainted with many-valued logic said, "Okay, he's just introduced a third value!" No, he hadn't introduced any ordinary third value! His third value was in fact not the result of an operation but one that referred to the operator itself. He called that a "rejection-value," a rejection-operator.

He claimed, "If I've got a proposition like 'Today the sun is shining,' then we've got to find out whether it's true or false." As an empiricist I look out the window to see if the sun is shining. If I am a logician, then I take down the great book of books in which all true propositions are contained and look to see whether the proposition "Today the sun is shining" is in the book. If yes, the proposition is true, and if it is not there, then it is false. Now, Gotthard pointed out that first I have to have a place into which I can put the proposition, "Today the sun is shining." And so I have to "input" the proposition *P* into something that is, as a manner of speaking, not there yet. But if I have a place, then I can assign the proposition *P* to it. First I have to generate a place, then I can anchor a proposition *P* there. If, however, I do not allow the place and reject it, then I also cannot put any propositions there.

Thus the task of the revolutionaries is to reject the proposition "King" right from the start and concede no place to it. Then the king just can't happen, he is neither low nor high, neither venerable nor detestable—there is no place for a king. I'm always telling my friends the revolutionaries, "You must study Gotthard Günther so that you can eliminate the king even as a 'yes' or 'no'—only then will you perhaps be able to create a world that has no more places for kings." The slogan is not "No king," but rather Wittgenstein's view that a proposition *P* and its negation are talking about the same thing. Therein lies a very important realization, in my opinion: If I want to talk about something, then I also need to have a place, a logical space, in which I can set this proposition. Therefore I find a logic, a place-value logic according to Günther, very important. For such a logic also proves useful when we're talking about life and about self-reflexive operators. If I have a place-value logic, then I can ask myself what the logic of the places in which I set my propositions looks like. Thus, from the beginning, I am moving in a second-order problematic.

Another related problem-area is that of operators for operators, particularly as this game could be pushed further: operators for operators for operators, and so on.

Yes, definitely. You touch on an important problem, especially as I have sometimes been a little unhappy with myself and my choice of concepts. After I had introduced the concept of the second order, from time to time in discussion the question would arise, "And what does the third order consist of, and the fourth, the fifth, and so on?" I could have avoided such questions if I had closed the circle right from the start. With the second order I was aiming at a reflection of a reflection—and not at orders to which integer index numbers are attached and which could be counted off far into the distance. We only need two forms; there is no third.

The second order of the second order is still the second order.

Exactly, that is the idea. But unfortunately this point is obscured by the number two. It is not anchored in semantics—and many hope to reach an nth level . . .

In the manner of the Russell-Whitehead logical types.[27]

Exactly. And with that we're already back again in the hierarchy of values or in the doctrine of logical types—altogether a sad landscape, because there we are hopelessly lost.

Let's conclude this one essential point of operators and the diversity of operators, and let's go on to the next, very important characteristic of living systems. They have the levels of genotype and phenotype attributed to them, and at the moment there are virulent debates in biology about the relative weight of these two domains. How do you see this distinction between genotype and phenotype as regards the essential characteristics of living systems?

Yes, here many other, similar dichotomies can be invented, such as "nature versus nurture." On the one side: What do I bring with me onto the scales by way of genetic material—nature. On the other side: What do I learn over the course of my life? To what extent can I bequeath what I have learned—culture. With good dichotomies you can earn good money. I, on the other hand, always try to show that dichotomies, in a sense, are always claiming the same thing. The opposition of true and false, the dichotomy of learning and genetics or of phenotype and genotype—again and again, these dichotomies come down to the same thing. Why do I say "the same thing?" They claim that I can make a distinction between this and that form.

If I look at something another way round, I automatically find the other side. So it's like the statement that a proposition P and its negation are both talking about the same thing—or, a proposition P stays the same with inversion. They are "invariants," they both express the same thing, they require each other so that they can oppose each other. That leads me, or us, right back to our conversation. If you're always talking about the systems there—and I'm always hearing, "There is an operation, there is a genotype, there is this!"—then I'm constantly asking myself, "Why is he saying 'there'? What cognitive strivings are making him say 'there' and point 'there'?" The fact that we are talking about these dichotomies that you have introduced belongs to the apparatus of our social interactions, that is, we use our mutual presence to talk about the opposites this or that. What fascinates me is, "Why are we talking about these pairs of opposites—this or that?" Of course, I can always go on the offensive. One claims, "I am the representative of Phenotype-Culture-Nurture!" and the other, "I stand for Genotype-Nature-Nature!" And soon they've started a splendid fight. In these faculty quarrels I always come out like this: "Ladies and gentlemen! Please sit down, and let's talk it over!" They ask, "Why? We're talking about fundamentally different positions," and I answer, "No, we're talking about the same thing."

I don't know whether I'm expressing myself well enough, when I say that we're really always talking about the same thing, but if we invent a dichotomy, then this dichotomy splits into two seemingly different areas. There are of course dichotomies that do not split, in which the one part is the complement of the other. But noncomplementary splits, I would claim, are "incomplete"—and therein lies the problem that demands a detailed investigation. Why do I operate with such "incomplete" dichotomies? That is the question that interests me. "Why would a dichotomy that represents no complete complementarity be postulated?" This question captivates me.

Let's present this matter of complementarity by means of a trial. Although in biology there is a quarrel based on a dichotomy, the distinction between genotype/ phenotype may also be formulated in this way: There is Heinz von Foerster, who sits across from us and gesticulates and speaks and interacts with us. That is the phenotype description of Heinz. And then on the cellular level of Heinz von Foerster, there are billions and billions of genetic programs in cell nuclei that contain the instructions for how Heinz von Foerster should be produced under suitable circumstances. With this we find ourselves on the level of a genotypical Heinz description.

Let me say it once again: I'm delighted by your description. And I was also delighted when I read the first papers by Lynn Margulis on the emergence of cellular nuclei and genetic programs. Her ideas on eukaryotic cells were very centrally concerned with symbiosis.[28] But we're still moving within the question, "What is 'there'?" In my game, I say that here and now, if you like you can draw distinctions so that genetics and phenomenology are considered to be separate areas. Here, however, language becomes a trap that drags me further and further into such separations, instead of preventing one side of the separation being made responsible for the other, or the other for the one.

Heinz von Foerster doesn't emerge, doesn't speak, doesn't eat, doesn't live, doesn't breathe because he is a phenotype but because he is Heinz von Foerster. This thing is a compact and self-contained affair whose hands now tremble, who raises and then lowers his voice. To make some part of Heinz von Foerster responsible for this shakiness of the hands and the peculiarity of the voice, and therefore for his moving like this and talking this way—this is the game of the observer, who needs to satisfy himself as to why this Heinz gesticulates so wildly with his hands. I can only look back to Gregory Bateson and his daughter's lovely question: "Why do the French wave their hands around like that when they talk?"[29] That fits precisely into this category. To this question you can answer, "They can't help themselves, they're French"; or, "It has to be that way because their locomotive-apparatus requires it"; or, "Why, that's just the French gene pool from which they all spring"; or, "No, the pragmatics of language require the French to wave their hands—they cannot speak with the glottis alone but must simultaneously speak with their hands!" So we could entertain ourselves with these answers—and then I get back onto my form platform and ask, "Which form of answer do we want to accept on the phenomenology that we're discussing?" So then we can go home happy and read the *New York Times*.

But in your work on molecular ethology, you also deal with the type of connections between molecular-genetic areas and types of behavior.

Yes, but the article, if you remember, begins with a critique of language. The nice thing here, the advantage that geneticists have in this discussion, is that they work with substantives and therefore are able to go back to a hierarchy, to a hierarchical structure that makes one noun explicable through another. If I am talking about problematics of behavior, however, then we are always using verbs. And verbs don't allow themselves to be ordered in a hierarchical system. Paul Weston and I conducted a study on the organization of word-chains. If you look in a dictionary, you ask yourself, "I'd like to know"—this was our example—"what is a pheasant?" "A pheasant is a bird." "Thank you, but what is a bird?" "A bird is an animal." "Wonderful, but what is an animal?" "An animal is a living being." "Good. And what is a living being?" In the end the dictionary always comes to the conclusion that it is a being, an object. The claims that the dictionary makes on you are, "Mind that you speak in English; mind that you see something; mind that something is there because we see it." And in conclusion, "Why are you asking me these senseless questions anyway?" Ultimately, that's what a dictionary has to say about it.

Now we want to try it with "to go," "to jump," "to run," and "to talk." It turns out that such a hierarchical orientation simply isn't possible with verbs. With verbs we are dealing with a semantic network—and that cannot be represented through a hierarchical organization. Thus behavior is a different logical type than the objects that appear in behavior. The entire explanatory schema, if we want to talk about the explanatory schema of behaviors and objects, is fundamentally different. Thus at the time I pointed out in the article on molecular ethology that the behavioral psychologists have great difficulty in falling back on such elementary concepts in the manner of geneticists who can develop a computer program for every area. That was the beginning—and as you see, it was a critique of language.

In the course of this article I tried to maintain this critique of language and to say: Be careful now, if you talk about such and such an area, then we must bear in mind that we are talking: We are moving within a framework, a structure of relations that has these characteristics on this side and those characteristics on the other side, so that we cannot go into the explanation-problem of the one with the explanatory schema of the other. What really amused me was the remark that we were trying to make the world into a trivial machine. The behavioral psychologists, for example, are busily working to transform us into trivial machines—they would so like to be able to explain us. And these remarks were terrific fun for Ivan Illich. After he had read the article, he rang me: "Do you know that you've ruined Skinner?" Of course I knew that.

Does my form argument make some kind of sense? Would this be the time to talk about this problem a little more? It's always possible to take something from it; for example, ask "What is life?" or "What is consciousness?" and I will explain the problem of form further. I could imagine that a reader reading this might simply have no idea what we're talking about. Now we have a form problem. And this form problem is as follows: there are fascinating questions that have a very specific form. These questions require, in whatever form they are posed, a form of answer commensurable with the form of the question.

The form of the question is very open, I claim, because we are entering a semantic area that stretches out its pseudopods or semantic tentacles in a hundred directions. By "pseudopods" I mean that a word is linked into a semantic area because the pseudopods in this semantic space reach out and start playing in a hundred directions. Now, if I'm taking this point seriously, I will ask, "If our semantic pseudopods operate in this particular way, how can I then keep our conversation about life going so that it will stay alive?" Because then I have—life. We could also stop our conversation and say, "Look in the dictionary if you want to know what life is." "Flip through the dictionary, then you'll know what consciousness is." "Turn to Shakespeare, then you will learn about life," and so on.

I hope it's clear that in this way I am trying, constantly trying, to keep this conversation, our dialog, going. That means I'm trying to work the pseudopods in my conversational partners' semantic areas, get them going so that they keep coming back to my pseudopods and so that in this way we—and this is the important thing—keep our conversation alive.

I would like to bring in an amusing example: Someone was giving a lecture about memory-power: "I will show you a series of meaningless syllables, nonsense syllables, on the screen, such as *vam, ku, pip, bab, lif, fem,* and so on. If I project this series onto the screen, it will be impossible for you to remember what it is going on; every meaning is excluded." Then I stood up: "One moment, please." I walked up and pointed to a point on the empty screen: "If you could just project that slide once more, just there, where my finger is—*fem.*" The slide appeared—and there was *fem.* "For me, this *fem* is a gigantic operator: I think of all the women that I have loved; I think of all the women it is possible to love. I see them naked and clothed. If you say to me that *fem* is a meaningless operator, I'll counter that *fem* is the richest operator that you could find. Thus I will never forget *fem.*" Everyone laughed, of course—and I played the fool for the auditorium. But it's precisely this point, the semantic tentacles of expressions, even of apparently meaningless words, that is so essential for me. During conversation, a peculiar dynamic arises through these mutual tentacular linkages. It goes "boom," the thing touches me; "Wow," on the other side. . . . Many people manage to stun these pseudopods. Education, work, a trivialization of everyday language all play their part—and soon people just can't turn somersaults in their own semantic spaces. My educational principle would be as follows: "Let them cut capers in these spaces so that they are constantly touching other relations."

At BCL we had the idea of building a machine that could talk with its partner and that was built out of semantic cultures, in which one might say that every concept forms a very large computer, which is connected with other computers: a computer of computers that form relationships with each other and that develop "clouds of connections" in these semantic spaces. "Green," "pot," "plant," but also "*fem*"—these are already fascinating operators and computers, even taken by themselves and they are capable of incredible things: they "know" how to stretch out their pseudopods and are able to point toward other things. But this is how we are able to talk to each other in a lively way. That would be my form of description for life and how I would like to play into and along with it.

One characteristic of living systems, our conversation included, is that on the basis of their internal possibilities and their repertoires, they are in a position to trivialize their surroundings. Is that a fitting continuation of our tentacle dialog?

Only under these conditions may predictions be made, and therefore explanations given, yes. But in this conversation I'm always wanting to point out that we're talking about living systems here and now. I'm always having to stress, "It is a problem of form," or "It is not a problem of form." It is the manner in which we speak about these questions that makes the difference. I'm always wanting to draw attention to the fact that what interests me about the question "What is life?" is not the definition that someone, you or I, conjures up. On the contrary: If we talk about the problem of life here and now, then this becomes important: In which form does this conversation play out? I want to stress, over and over, that it all comes down to the conversational form we use when we talk about life. It isn't the problematic of life that fascinates me—of course it is one of the tremendously fascinating problems—but rather the form of speaking when we talk about life. This point fascinates me: Through the ways that we talk about life, we create, bring forth, produce life.

Through the dynamics of speech, through the dynamics of our being-together, through the dynamics of our conversations about life, life emerges. I want to point this out again and again because in some cases one might just say, "This explanation is unclear to me," or—to stress even more strongly this incredible thing, that my grunts and sibilants engender grunts and sibilants in my neighbor: "Thank you, now I know it," or, "Wonderful, now I have understood it"—or maybe even, "Dear Heinz, you still haven't told me anything about life."

When we are conducting our dialogue on life, we are able to locate it between biology and biodialogy. And there is a really beautiful Foersteran quote on biodialogy: "The laws of Nature are written by man. The laws of Biology must write themselves."[30]

I'm very pleased to have written that once.

This proposition is terribly interesting and far-reaching, especially in the sense of these pseudopods or the semantic tentacles. What can this proposition mean, from the hundreds of possibilities, that life writes its laws itself?

For that I must return once more to this fascinating phenomenon of eigenvalues and eigenbehaviors. What's astonishing about it—since we are in the field of language—is that laws can be turned out in such a way that they write themselves and in the end we are able to talk about the "laws of life." Through such recursive operations it becomes possible for eigenbehaviors to peel themselves out of infinitely many possible representations. That means the problem of talking about life becomes an eigenproblem. But how can I find a proposition that finds itself? It is in precisely this form that I see the problematic of talking about life. And that's the joke, saying that the laws of life, the laws of biology, must write themselves. Now of course the next question is, "How do you write such laws?" And, "How can you write them?" In this way we set off on the search for a description that describes itself. The problem of life does not primarily consist of fixing criteria for determining, "There is life, that is a living organization." For me it is incomparably more urgent to find a description that finds itself, or a description that writes itself. And so not, "There is the law ABC," but rather, "Here is the law that has set itself." And if now I ought to conjure up this law—"Heinz, produce this life-law that writes itself"—then I answer, "I still haven't found it."

Of course I could say, "Here I am."

Yes, one could say it like that. It swims from one form to another.

What's interesting, and it is also a change of one form to another, is the opposition between "life laws" and the "nature laws" that are written by men. In one case, that of nature, we have man as "observer." In the laws of biology, if they write themselves, then the observer seems to be eliminated.

No, he is co-created. Even now he is there, although neither explicit nor implicit. I don't say "there," but rather, "Here it is." That's why I found it really lovely that you said, "Here I am."

Normally biology is considered to be natural science or a very important branch of natural science. And yet for you there is this distinction between the laws of nature, which are invented, and the laws of biology, which write themselves?

It depends which direction you're coming from. If you are coming from the natural-scientific direction, then you are already quite happy to say, "The worms divide themselves into . . ." "The living creatures divide themselves into . . ." We already had that lovely distinction of phenotype and genotype. That is the version of biology dominated by natural science, to which physics and astronomy also belong.

If, however, I want to consider biology as a science of life, then I deliberately do not say, "There is life," but instead pose the question, "How do I build myself into life so that I become a co-describer of life by describing myself?" then the categories and the forms of speaking about biology become fundamentally different. Of course, I can always retreat to physics or the description of something else entirely. And then once more it is I who determine the laws of biology—and they depend on the observer once again. My invitation in the proposition "The laws of biology must write themselves" is, however, "Couldn't we and shouldn't we find an eigendescription that is structured so that it describes itself?" That is the eigenvalue problem of life.

THIRD DAY

Movement, Species, Recursion, Selectivity

The more profound the problem that is ignored, the greater are the chances for fame and success.
—HEINZ VON FOERSTER, *Understanding Understanding*

Seemingly, I am performing "thought-experiments." Well, they're simply not experiments. Calculations would be much closer.
—LUDWIG WITTGENSTEIN, *Last Writings in the Philosophy of Psychology*

And God said, Let the waters vnder the heauen be gathered together vnto one place, and let the dry land appeare: and it was so. And God called the drie land, Earth, and the gathering together of the waters called hee, Seas: and God saw that it was good. . . . And the earth brought foorth grasse, and herbe yeelding seed after his kinde, and the tree yeelding fruit, whose seed was in it selfe, after his kinde: and God saw that it was good. And the euening and the morning were the third day.
—GENESIS 1:9–13

In the sense of a game with the living world in its simpler forms, we would like to turn first to the concept of recursion, then later to problems like movement, energy, growth, simpler forms of orientation.

I've already pointed this out several times: The whole problematic lies in the language. If we understood the problematic of language, then we would have the possibility of dealing with the problem, "There is a living being," "There is a galaxy," and so forth, in a manner that projects this problem onto language, the sayable.

Not long ago I reread an old paper that I gave in Royaumont at the invitation of Edgar Morin, "Notes on an Epistemology of Living Things."[1] It delighted me that even back then I emphasized in the first two points: The most pressing problem is language—even if you won't find it in exactly this form. At the time I formulated the problem as a problem of "representation"—I wanted to use the word "description" for a so-called final representation that finds its expression motorically. Back then I distinguished between several processes, between representations of experiences and the experiences themselves. After many manipulations and operations they finally yield a final representation, a terminal representation that I called description.

I suggest we take a big step back so that my hope of recognizing the problematic within description doesn't have to be abandoned. We should move away from the question "What is there?" In the proposition "Here is a comet,"[2] we should draw the attention away from the comet toward the "is." If I say, "There is a comet!" one just doesn't hear the "is," that vanishes under the rug. Let's look at the "is," and then we could look at the comet as well as at Karl Müller, Albert Müller, Heinz von Foerster, California.

Let's turn to the "is." There is, for reasons that we already discussed in our first conversation, almost a magic path, namely recursive descriptions and formalism's recursive functions. As a starting value I would just like to ask you, what do this magic and the strange attractive qualities of recursive formalism consist of?

I've got to admit that in the beginning, I had troubles with recursivity. At the start I couldn't make anything of it. That means that at first I only saw that it creates logical problems, problems for the calculation of results if a description contains itself. I said to myself, "If the described is in the description, that's a logical sleeping dog. One should let sleeping dogs lie. There is a loop there, but you don't get any further, that is the 'end of history.'" So I thought at the time. Then in mathematics I encountered functions characterized by the unknown coming back in the description of the unknown. There are very specific mathematical forms in which such phenomena occur. I looked at certain functions; I played with them, worked through them, experimented with numerical magnitudes, calculated and calculated and calculated.

At this point I did not yet possess a computer. At the time not everybody owned a little pocket calculator or a PC. I did these calculations with slide rule and fingers but didn't get very far because the precision of the slide rule simply wasn't sufficient. You could see out to the second decimal place but not the third. And suddenly I got my first calculator—that must have been 1968 or 1969—a Hewlett Packard, a tiny apparatus, a magical machine. I took this calculator and said, "Well, brilliant! Now I can write the program, 'one, two, three,' and now I can finally get the whole twenty or a hundred decimal places." And actually, I got astonishing results; the recursions converged. For the first time I saw the phenomenon of "eigenbehavior." Then I came back to the great mathematical formulas of David Hilbert.[3] If the derivation of a function results in this function again and one might say it describes itself—in 1895, Hilbert called that the eigenvalue of a function.

I see these Hilbertean designs: "My God, eigenvalues come out, good old, seventy year-old Hilbertean eigenvalues!" I was delighted; I kept playing, kept calculating and spoke about these eigenvalue-observations on several occasions. I can tell you the first function in which I discovered this overwhelming game of recursions and that was such a big surprise for me. Take the following formula: What is x to the x to the x to the x to the x to the x, ad infinitum? What comes out if I take any value for x, what happens then? Under certain circumstances one can find an eigenvalue for this operation—and the funny thing about it is that you can see it immediately. If you know that an eigenvalue is going to come out, then the endless series of x to the x to the x to the x represents the eigenvalue you're looking for. If I substitute a y for the endless cascade of x's, then I can, for example, write

$$x^y = y$$

Because over the first x there is this endless x-series that is expressed with y. I can immediately solve this equation for y—and I've found my eigenvalue. I've just got to go through infinity and back; then I've got the answer.

Expressed differently, one needn't perform infinitely long calculations because the infinite calculations are represented through their eigenvalues. If one sees that, then of course it just gets more and more fun. Infinite recursions are thus not unsolvable and the "end of history" as we believed earlier; infinite recursions lead to a stability or to several stabilities—that, of course, is earth-shaking.

What do such eigenoperations mean, however they might be built? Perhaps that we shake hands with each other or that we say "Good morning" to each other? They mean that a stability has peeled itself out, because "Good morning" represents an equilibrium between the two of us. I say "Good morning"; you answer, "Good morning." We both know what is meant by "Good morning." Or, better yet: We don't need to know what is meant by it. We know that "Good morning" functions like that. It creates itself. I came upon all these new phenomena through this little pocket calculator. I owe it the intuition of recursions. The physicist and mathematician Feigenbaum also describes a similar calculator experience. With the program that he entered into his little calculator, he could suddenly gain insights that had previously been closed to him.[4] The trivial calculators of those days proved to be invaluable teaching materials.

For better understanding, I'd like to bring in two examples. The first is an example that you use all the time as well, that of the root-finder as operator, which leads to the eigenvalue "one," and the second is the operation "add one," which goes to infinity without producing eigenvalues.

The eigenvalue of this addition is simply infinite—we just aren't able to do anything with it. It brings no stability, and it doesn't converge. Certain operators diverge, and my claim or my suggestion is this, that the operators that diverge simply vanish. Only those operators that lead to stability stay with us. Everything that isn't stable explodes, you might say, or eliminates itself, doesn't appear any more, cannot come up any more, vanishes. Therefore, we can see only the stable operations. But if we're only ever able to observe the stable processes, then we're always trying to ask, "What's behind the operations that have produced Karl Müller?" And this is where my important point starts: We can never know that, although we can know how we should treat Karl Müller, although we have concluded, "If you one says 'Good morning,' he responds, 'Good morning'; to the question, 'How are you?' he replies, 'I'm fine thanks'—thus far everything's working wonderfully." But now I want to know "Why?" I want to know, in the classical sense, "Can I set this mechanism on the table and claim, 'This mechanism exactly determines Karl Müller, who answers, 'I'm fine thanks,' if I ask him, 'How are you?'"—I claim that I cannot determine this mechanism. Although I can set a mechanism on the table and ask it, "How are you?" and it answers, "I'm fine, thanks," if I ask it something else, such as "Tell me, what do you really think of Richard Wagner?" well, then one would first have to program the mechanism so that it would either say, "Nothing," or, "He is my God," or maybe, "He is a dangerous seducer."

Now we have as an example Ashby's "evolutionary operator."[5] You have a series of numbers. Take two numbers, multiply them, substitute both numbers for their products, and so on. After a longer sequence, it leads to odd numbers, and finally there are only zeros left. We can, however, say how this operator works or why it leads to zeros.

But you cannot deduce the operator from the zero. From what we see we are always wanting to deduce what produced it. We should give up this explanation urge; rather, we should ask: If I see how it is, how will it go on if I get it involved in a further iteration? The hope of explaining why it operates this way and not otherwise I leave to people who believe that they can find out why it does that. At the same time I take my own quiet pleasure in it, because with every such "explanation" I know more about who the explainer is.

Recursive descriptions can be used in the whole casino of the world, from plants to animals and social behaviors right up to our conversation.

Would you now like to speak from the position of plants, or would you like to speak from the position of Karl Müller who talks about plants?

From the position of the questioner who asks Heinz von Foerster . . .

What he thinks about plants?

. . . what he thinks about recursive descriptions with regard to plants, animals, people, and conversations.

We're already much too far ahead. Let's take a step back. First, we need to pose the question, "Why can we use the word 'plant' at all? Why do you understand if I say 'ierhg,' not 'hufg' or 'klar.txt'? There we've got the first problem. If we manage to come to terms with that, then the next one comes up at once: Why should I use precisely the expression 'plants'? Why not 'dance,' 'chance,' 'romance,' or something similar?" There are cascades and cascades of problems that—and now recursive formalism gets into the game—come down to this: Whatever I claim, whatever emerges as stability in these speech-cascades—that "plants" means something to both of us so that we could both come up with a suitable example or point to a relevant object—that is based in our special social modes of behavior. In our social behavior, the dialogic is organized as a recursive process in which we don't know what we're talking about as long as an eigenvalue or stability hasn't emerged from this recursion and settled itself. Only after some time, when these stabilities have established themselves in a society, do we know what is meant by the funny sounds that produce the word "plant" and with which actions, operations, or doings they can be connected or woven. In our constant life-with-each-other, a recursivity of being-together emerges: A talks to B, B to A, A to C, C to B, D to A. In this recursivity a common language emerges.

Then semantics owes itself to such recursive practices?

Yes.

Let's talk once more about this recursive formalism and its characteristics. I think an important point for commentary is that by this time with regards to terminology there exist the most various expressions for these phenomena, coexisting peacefully, or less peacefully. One variant is "attractors," another is "states of equilibrium." Attractors take on different forms, point attractors, threshold cycles, turn themselves into strange attractors, and so on.[6] Can you explain what in this area is equivalent to the terminology of eigenvalues?

For me, of course, it's ridiculously funny that the people who are dealing with recursive functions, with eigenvalues, which have been called "eigenvalues" for a hundred years, have renamed them. I don't know whether they have any idea that these games were already being played at the end of the nineteenth century by mathematicians like Koch, Poincaré,

Hilbert, and so on, with a nomenclature derived directly from mathematics. My suspicions are in the direction—that they knew nothing about it. I've looked at the ages of these people who have developed this new terminology; they are all around thirty, forty, or fifty years younger than me. Thus I can well imagine that they just don't stumble upon these old concepts anymore. At BCL, from a mathematical perspective, we were dealing with very similar or analogous problems, without, however, using a vocabulary like "chaos theory" and "attractors." But for us the fun lay in observing the convergences of certain functions.

Interestingly, this attractor-nomenclature is composed of strict anti-cyberneticians. They held cyberneticians to be teleologists and alleged that they were always asking about the "where to," about the "target." That precisely these people let teleology in again through the back door in the form of "attractors" will always baffle me. I doubled up with laughter: Now here comes the "chaos nomenclature" with an "attractor" that pulls—why not with a "pushor" that pushes? That would be the more correct formulation. Nothing is attracted or pulled; it's constant pushing and shoving. But this is the material from which the paradoxical history of science will be built. I have, however, concluded: My young colleagues have reinvented what Mr. Koch had already seen in 1895. If one divides certain configurations over and over again, then something peculiar comes out of it.[7]

Benoit Mandelbrot in particular has "discovered" a lot in this area—the "thing in itself" was already existent—and now one speaks of "fractals," "fractal dimensions," and so on.[8] But even that had already been dealt with by Koch at the turn of the century.[9] The new thing about it is the machines: With the machines you can create operations that Koch couldn't do with his slide rule and pencil and paper. And the fun that people like Mandelbrot had, a fun they never divulge, consists simply of this: They generate their figures with complex numbers; they play on the complex number plane. Here the points no longer lie on a line but rather in a plane—and now they can let these points whizz about on the plane—and in this manner the most amusing forms are produced. The great game consists of generating a suitable computer program, but that is relatively easy. Peitgen and another group, for example, came upon the idea that we at BCL had also already developed: If you place a monitor in front of a video camera, then one can visualize recursive functions in a simple way.[10] You begin with any signal, and soon this signal runs in a peculiar manner. One time it produces a gorgeous square, another time pulsating figures, etc. Eigenvalues can thus be demonstrated with a video-loop that takes the output back in as input. And that was also done with fractals and was widespread, not in the little BCL lab but with lots of research funding and public resonance. But of course that was only thirty, forty years later—and everybody already understood what comes out of it.

The technological, computer-oriented element here provides very strong support for the idea that the ways of speaking and modeling from "dynamic systems" were able to spread quickly because programs could be created, experiments and simulations could be carried out. In Toulmin there is the interesting observation that this lead to a "paradigm shift." Previously one knew of these phenomena— keyword "Koch's triangles"—but at the time they represented a typical peripheral subject. Only with the technological foundation did the situation reverse radically: What had been peripheral moved to the center—and vice versa.[11]

I don't like the term "paradigm shift." Paradigms don't shift. People shift; humans change, but paradigms don't change. They've got no idea how to change. Paradigm shift is a misformulation. Once more attention shifts to the paradigm rather than to the person who's talking about it.

If we turn back to our recursions, then there is a very interesting and important point: On one side there are operators that diverge, move apart, explode; on the other side are those that converge. Don't convergent operators have an evolutionary advantage over the divergent ones?

It's not just an advantage. Divergent operators disappear; they cancel themselves out. This form simply cannot be sustained. They run for some time—till they vanish. Of course now one might think: Is the operator itself the result of a recursive meta-operation? And I claim that actually is the case. The stable operators, the stable eigenvalues are themselves eigenoperators of greater "meta-operators," and probably one could stop and dwell here on this second plane.

Main subject: "world as a casino." If one takes a simple calculator, there are two points of interaction concerning this game with the world; the first lies between the input—what comes into the machine from outside—and its inner conditions, and the second between the internal conditions and the output.

There are so many different ways of describing something. May I come back to language again? If we start playing a certain game with the world, we can only play it because we have already played lots of other games, and now we're saying, "Now we want to act as if there were a world and behave as if this or that were the case." So we could claim, "Right now we're a school class," "Right now we're a club," "Right now we're in a conversational phase that's gotten stuck in the foreplay, in the prologue." First of all we want to actually find out why we can talk about a world, why about an input, why about an output? If at this point I accept your suggestion, then we're already behaving as if there were a world, as if certain interfaces existed, and so on. We have already accepted too many presuppositions for my taste to be able to play the game.

If I were in Vienna in the 1930s, then I'd say, let's go to Mühlhauser, the big toyshop on Kärtnerstrasse. We go in: "I'd like to buy an entertaining game!" The sales assistant comes over, puts a game on the table, I buy it, and with my friends I start to play Monopoly. In the instructions it says under which circumstances I lose, when it's my turn, how to roll the dice, what the goal of the game is, and so on. After some time, however, this gets too boring for us, and we decide: Now we want to invent a lovely game that our children could play, that would keep them happy and keep them busy in a stimulating way—without fighting or teasing each other. For this purpose we want to invent the "Game of the World" or "Casino World," or buy them at Mühlhauser; they have the biggest selection. What do we need for it? Here we are in the prologue, prelude or foreplay—and this in every sense, including in the musical sense of an opera.

I'm asking myself whether the conversation that we've conducted so far has already a provided sufficient prelude so that one could now go purchase the "Game of the World" or "Casino World" at Mühlhauser, without those who, by reading play along, believing that we are now dealing with the only game available for purchase at Mühlhauser? Do you understand my question? I think that this prelude, which is leading to the purchase of the game, is still incomplete. If we start going deeper into the game already, then the reader who is purchasing this game with us could object, "Why are you even going into this Casino World?" or, "Why 'world,' 'I,' 'you,' 'game'?" Or even, "Why at 'Mühlhauser,' why 'purchase'?"

We're trying to develop such games right now—we're in the middle of the creation of "life."

I can imagine that the reader would be interested in what's happening at this point. We've talked about living systems. We haven't talked, though, about tealeaves and blossoms; somehow it was always the level of animals implied, not of plants. If I were a good Maturana, then I could develop the identity and the differences of plants and animals, whereby the one area goes around with a chemical system or with a chemical interactive system and the other with a nervous system, and so on. If you asked him, Maturana could develop that beautifully. I, however, am unfortunately incapable of that because I don't know enough chemistry.

While we're at it, I'd like to seize on another point: It is extremely interesting to see that we possess a language in which the development of different units within a changeable environment can be described. At this point I'd like to introduce the idea of a distinction. And the moment I introduce the idea of a distinction, there must also be a mechanism that can make a distinction. That means there exists the possibility of being able to distinguish an A and a B from one another. In this case certain symmetries also arise that I don't want to develop more closely right now, but the prerequisite "calculation of a distinction" must be given, otherwise I cannot stipulate a distinction between a unit and an environment. This stipulation is already based on the ability to draw such a distinction.

Assuming that right now we're not interested in how something like that comes to be, rather we're just saying, "Good, I believe you, Heinz, I believe you, Karl, I believe you, Albert, that one can invent something that can draw such distinctions." We're postulating the existence of such a great distinguisher—and we're able to begin speaking about a unit and its surroundings. But we're also put in a position to pose further questions: "What are the interplays like between this unit and its surroundings?" If I want to write or say something about that then it turns out like this: We're not confronted with the operation of distinguishing but rather with the results of the operation of distinguishing. That means what I want to direct attention toward is the issue that it is not the distinction itself but always the result of a distinction, that which I at first, rather unfortunately, called a "representation."

That's an important point for me. Because if one does not separate a distinction and the "representation" of a distinction, then without realizing it, one manages to get into a very nasty channel with hardly any chance to turn around. One can start the following game: These are my grandmother's glasses. We must now be able to distinguish "These are my grandmother's glasses" from the operator that has distinguished my grandmother's glasses. These are not, however, the grandmother's glasses meant by some theoreticians of memory when they go off looking for a little picture of my grandmother's glasses in the head and "memory" of the grandson.

All of these steps would be necessary for buying our box with the title "Game of the World," "Dance with the World," "Nature and Society Game," "Games for All Ages," and so forth. And after all these intermediary steps were explained then we could ask, "In what ways does the dance, the game, the interaction with the world take place?" Before that I would need a special distinguisher, a "fine distinctor," something that is able to operatively draw functioning distinctions. If I look at my grandmother and see her glasses, then I can and must talk about her glasses because otherwise I would not have seen them. We see very quickly how strange these dances, games, and interactions with the world are. Because it's to do with games, between Karl, Heinz, and Albert there are games for three, trio dances, trialogs, always in a social compound. Only once these prerequisites are given can I set off in the direction of Mühlhauser's. I buy myself the game box—end of my digressions.

If we open this box, what do we come upon first?

First of all, we find a big piece of paper on which the inscription "World" is printed in bold letters. "Ah, there's a little round disc moving there—that's Karl; here are two little figures—one is Albert, the other Heinz. Now these three playing pieces are starting to creep around the board. And one says to the others, 'Have you met Albert yet?' 'No, where is he?' 'Just keep going!'" Boom, the two bump into each other: "Ah, there's Albert!" In this way, the games with the world develop. Now you could say, "Now let's explore other places on the big piece of cardboard, on the pasteboard that says 'World' on it." Well, then the three of us are walking along: "What's this? Hmm, lovely palm trees, the sea: Florida! Let's go swimming, check out some girls, play ball, whatever."

This game plan leads directly to today's quote from the Foersterian fundus: "(Movement) → (Change in sensation), but not necessarily, (Change in sensation) → (Movement)."[12] *The big sheet that says "World" also makes it possible for us not to have to visit Florida; we could also stay here and now . . .*

. . . talk about it in California, yes. But watch out, with these theorems one must take another big step back again. The step back looks like this: A great endeavor that I undertook for many years was the attempt to free the world from the idea of the "depiction."[13] On this point I was, incidentally, in complete agreement with Humberto Maturana, who also wanted to free the world from the idea of its depiction.[14]

Freed from the depiction, but not from the "representation." If one wanted to do that, one would have to at least talk about presentation and not about representation.

Absolutely right. I know, the word "representation" wasn't well chosen back then. When I reread my article, it gave me a bit of a start to see that I'd used the word "representation." I wondered why I hadn't used "description." Later, however, I saw that description is a special form of representation. Representation should by no means be taken as a repeated presentation or a counterpresentation. Sometimes over the course of one's life one slowly gets a clearer idea of what one had intended earlier but hadn't yet been able to achieve because of a vocabulary hanging round one's neck like a heavy weight. By now it is completely clear to me that I could and should have said "presentation." "Free the world from the depiction"—at any rate I wanted to achieve that on two levels: First, in that I do not "portray" that tree over there inside of me; and second, in that a painter who paints a picture "portrays" nothing—the problem of art is not a problem of depiction but rather a problem of the observer. That means that once again it is not the speaker but the listener who determines the meaning. Thus if it could be managed to force the terminology of depiction, the terminology of representation into a quick and orderly retreat, various areas would become easier to distinguish. My language, that is one field, these trees represent another area, the trees that I am experiencing here and now, yet another domain—the wheel of depiction doesn't roll between them.[15]

In our game box, under the big board there are countless apparatuses, with or from which one can learn . . .

. . . yet again my theorem of learning: "One can learn even from the dumbest"—why not from little apparatuses . . .

. . . for example, from those in which one can enter an input and from which one will get certain output values with a beautiful regularity that one can predict—for example, from a computer that "instinctively" multiplies every number by two. Then I see various little apparatuses with several or many inner conditions that we can no longer calculate from the outside. They are all important teaching tools for us in our games with the world, our dances with the world, as you like to say—but also for learning to understand ourselves better.

I see what you want to ask. There are different dances: the shimmy, the tango, the foxtrot, the waltz, and so on. Are there any teaching tools that will get me waltzing faster than if I call on a non-waltzer and invent the waltz together with him? Well, of course there are; there are all kinds of possible aids at our disposal. Should I now say, "Yes, we do have such possibilities, and I'm glad to say that such apparatuses can now be purchased cheaply at the market"?

You yourself have contributed a great deal to the spread and the greater afford-ability of such machines.

I've actually built the type of machine you mention quite often with my friends in order to draw attention to certain possibilities, but also to particular difficulties. You have just described a machine that always multiplies by two. You will quickly figure out that and how it performs this operation. You put in a three, it gives you six right away—and if you know the numerical series then you also know, "Before me I have a two-multiplier that can't do anything else." Once you've interacted or danced with this apparatus a hundred times and the same thing has always happened, you might long for some change. It might also be that you go into a store with a modest wish and say, "Dear Mr. Mühlhauser, I need a machine that will always multiply by two!" "With pleasure, I hereby give you a guarantee that it will multiply by two for the next thousand years as long as you use the right plugs." Here we're dealing with a trivial machine. In our current learning context that means: I give this tool to your little boy as a present, you lose the instructions—and very quickly find out again how this machine operates. In fact, the analytical problem is solved in five minutes.

With nontrivial learning and teaching tools an ingenious idea comes to my aid: Let us, machine and man, form an informative as well as operational closed system so that the machine and I are moving in the same area. And suddenly we're able to start moving into an eigenbehavior, into eigenvalues that are also so kind as to remain stable. The consequences for our machine learning are naturally enormous. With nontrivial machines you should say to your son, "Dear boy, if you want to find out how the machine works, then you'll have to study from the beginning to the end of the universe. And when the end of the world comes, you still won't have found the solution. Enjoy the possession of this machine; play with it as long as you have it." You yourself know only that if this machine gets into a circle it will stay in this circle. That is the only thing that you can know about this machine. If it gets into the circle—seven, three, one, four—seven, three, one, four—then you can impress your boy: "Now it's on seven, just watch, the next number will be three, beep, three appears." And if the little one stubbornly keeps asking, "But Daddy, why?" then you'll give him a smack.

With that the learning and teaching machinery take on unexpected additional dimensions.

You see I have enthusiastically adopted the suggestion of learning and teaching tools.

There are, as with all these game kits, expansion kits also, and most of all instructions for how one can put together more of these machines. According to the directive from our first conversation, "More means diversity," all kinds of surprises would make an appearance.

The fundamental theorem of the composition of nontrivial machines shows: A composition of nontrivial machines is to be treated as a nontrivial machine. You can thus sit down at any point in the network of nontrivial machines, cut through it, anchor yourself into this net and can treat the whole system as a single nontrivial system. Of course, these kinds of ensembles become complicated, more complicated, and yet more complicated, but that doesn't mean that it becomes more difficult for you to figure out its way of moving. It will be as difficult—or impossible—for you as in the case of elementary systems. It's just that the assessments in the area of circularities won't be so easy to make any longer—seven, three, one, four—seven, three, one, four. In such configurations, so many additional factors play a role.

We've discussed extension kits using the plug principle. Then for the advanced levels there are also building kits with which operators of operators can be constructed.

Therein lies a second area that we briefly touched on earlier. Let's take an operator that sets certain actions, and let's imagine that it is also an operator that is kind enough to produce eigenvalues and therefore runs into one or more stable modes of behavior. The question then is, "Where does this operator come from?" Could we think up something whereby this operator itself becomes the result of a recursive meta-operator?

And indeed—and therein lies an important suggestion of mine to all psycho-, physio-, and socio-therapists—it's not just about eigenvalues; functions can also be brought into consideration. How should we understand that an operator represents an eigenoperator and that its results are subject to another higher operation? Well, the mathematical theory that lies at the bottom of this issue is called "functional calculus." The problems that one wants to solve lie in the functions, and the units being sought are called functors. A functor calculates functions or changes them. For example, one could explain to mathematically educated people: So-called differential calculus is a functor. Why? It assigns a certain function, let's say $y = x^2$, a second function that comes from the differentiation of x^2. And one learns nicely at school: A derivation can be written in form dx^2/dx; its result is $2x$; you just need to take the two from its position as an exponent and turn it into a coefficient. The differential operator transforms the function x^n to nx^{n-1}. The classification of these two functions x^n and nx^{n-1} is called a functor, and in our particular example this functor is just called the differential operator. One could, of course, imagine functors with much simpler dispositions—for example, the functor that multiplies by two. Here we're dealing with a functor into which one can stick any function, and on the other side out comes the same function—multiplied by two.

I recommend the following strategy for the study of those functions with which we're currently dealing: "Let's study the qualities of functors, since they are the results of that which we see happening before us." In one case I operate with a function, and its results lead to stable eigenvalues; in another case a functor operates on a certain function, and the results represent eigenfunctions. In this manner, one transfers the problem into those areas, psychology, sociology—the social sciences—that in my opinion oppose it at first. One needn't always think about numbers or functions. One can also talk about language; we can consider semantic operators; we can understand semantics as a functor that after a fashion establishes stable functions of reactions in us; and so on. For me as a describer, this means that I can talk about a semantic structure as if it were a functor that calculates, works out, and develops its operative stable functions and produces them through recursive functions so that I can reliably treat certain statements, claims, poems, and the like as stable functions. Thus, it doesn't move on the comparatively lower level of eigenvalues or eigenstatements or eigenwords, but on the next-higher level of eigenfunctions, eigeninterpretations, eigenrepresentations, eigenpresentations.

And this game may be played out even further with the construction kits?

The question is: "Do we need higher levels?" And I suspect that we don't need such things. I think that the functors can be positioned so that it is as if they produced themselves. I would like to develop this point somewhat more closely: For at this point the self-organization possesses the possibility of postulating an ensemble of self-operative functors because both the level of operative possibilities is so high and their richness is so great. In my little paper on the epistemology of living things, I have, by the way, already pointed out such a possibility.

If we're dealing with elementary forms of movement, then we must touch on a further point. In order to get these apparatuses that we find in our game box moving at all, we require energy.

If we're conducting these abstract discussions with units or systems or machines, then it is always already going to be the result of, as we emphasized yesterday, a "self-organization." And that is only possible if energy is already existent—otherwise nothing at all would happen: "Nothing comes from nothing," that's the first thermodynamic principle for our game box. Only the flow of energy allows the emergence of such units. And if this energy is pumped through the system, then very early on the possibility emerges for a system to change its surface. If the surface changes, then movement emerges. And on the basis of this abstract idea—movement corresponds to a change in the surface of a system— one can now actually take big jumps, little jumps, turn around in a circle, and so on. I will first take—so that you'll see why one should be careful—a big jump. If you remember the "Epistemology of Living Things," then a crucial point in that was that the change in the surface of a system first takes place linguistically as a change in the description of what has happened to a system.

That might at first glance appear to be too abstract an idea, but we can imagine this matter for ourselves roughly as follows: An ur-organism, a very simple, ten-celled organism, is floating in the water. It contracts under very particular circumstances, for example, if its immediate environment shows a special chemical configuration. With the help of its contractions the floating organism can move in the water. We might interpret it like this: "It finds the conditions too acidic," or, if you like, "It says to itself, 'It's too acidic for me here,' or, 'I want to get away from here.'" My claim at any rate is that the changing of the surface of an organism is an act of description. It can be an expression of an inner condition, an expression of something that an observer could note—an input, an irritation, a stimulus. If you are a zoologist, we'll talk about stimuli; if you're a physicist, we'll discuss causes; if you're a psychologist, then we'll speak of motives or even motivation; if you're an interpretive sociologist or a historian, then at this point you're given to talking about grounds.

Why are we talking about an organism at all? Because it represents an organized unit and from very early on this organized unit is able, through internal mechanisms, to change its surface. Naturally we can say, "But plants can't do it!" With a plant at first it looks as if it cannot move itself from the inside out. Nevertheless the inner conditions of plants are constantly changing. That which we see externally as movement in so-called animals is in plants hidden and covered by constant inner diffusion, metabolism, and so on.

On the level of plants one observes, among other phenomena, the orientation toward sunlight.

Of course, yes.

Also, if one considers a plant population, then it makes sense to talk about "movement" or "diffusion."

Absolutely, only in this case one sees the problem of movement realized slowly. But of course, here too there is constant movement talking place. Why do we need movement for our analysis at all? For me the aspect of movement first became central through Jean Piaget—and this wonderful work of Piaget's, *The Construction of Reality in the Child*,[16] which gave me, as it were, a moving experience of its own sort: as the saying goes, something dawned on me. It represented such a surprising insight for me when Piaget stressed that children control their reality by constantly playing with objects, and the objects with which they busy themselves change and keep moving. Suddenly I noticed the many linguistic metaphors like "to grasp," "to catch on," "to get it," and all kinds of others. Comprehension is also a motor activity—in French *comprendre*, in Latin *comprehendere*.[17] Clearly language is whispering to us, "If you want to grasp something, then you've got to grasp it!" That is also expressed in the word "object." An object is something that "objects" to me; I cannot make a movement, a gesture because something "objects" to me, and this obstacle is called an "object." These verbal hints and clues that language constantly and freely suggests are, however, for the most part overlooked, but for the first time Piaget brought me up against it: "Heinz, 'grasp,' 'catch on,' 'get it,' 'object,' 'get a handle on it,' do you understand that?" And I understood.

That fits with a Foersterian proposition: Movement → Change in sensation.

Yes, yes, exactly. I've asked myself how such insights could be deepened and generalized. And when I wrote my "Epistemology of Living Things," then I saw for the first time an analogy between movement and logic. How so? Well, our elementary ten-cell organism, which is still just drifting unconscious in the water, continuously floats between affirmation or negation. If an area displays too high an acidic concentration, our ten-celler pulls back with a "no." If this "ur-beast" is moving in an area with lower acidic concentrations, one could categorize that as a "yes." In this context I encountered a second, very interesting point about movement, which comes from Susan Langer. A philosopher who appeals to me greatly, Susan Langer wrote this beautiful book, *Philosophy in a New Key*.[18] There is a very funny passage in which she talks of so-called conditioned reflexes. The conditioned reflex works like this: You show a dog a piece of meat; it's happy to see its food and produces drool, its mouth waters—in technical language we'd say it salivates. At the same time, you ring a bell and then give the dog the meat. And this ritual you perform again and again: first the bell, then the meat. Finally you ring the bell without having the meat—the dog salivates nonetheless because it "believes" that the meat is coming now. Susan Langer says that in such situations lie the beginnings of "true" and "false"—not of "yes" and "no" but "true" and "false." If I ring and there's no meat, then that's a lie, then the ringing was a lie. If, on the other hand, meat appears at the ringing of the bell, then this is true because in this kind of dog's life, ringing means meat. Susan Langer argues that with the emergence of conditioned reflex the beginnings of "true" and "false" were also set. Thus, there are two logical forms. The first is "yes" and "no"—and proves to be independent of "true" and "false." The other deals with descriptions that may be "true" or "false." I found and still find this constellation exceedingly interesting.

At this point I'd like to bring in just one more footnote to the Pavlovian experiment because it will be important for further understanding. There was a Polish experimental psychologist by the name Konorski who repeated Pavlov's experiments. He was able to do this so well because Pavlov ran his laboratory trials according to highly meticulous protocols so that these experiments could be replicated exactly. One knew where the dog was held, where the assistant with the bell stood, in which direction the assistant was looking, and so on. Jerzy Konorski was thus in a position to repeat the Pavlovian experiment: the same equipment, the same dramaturgy of events, the same production. And actually, yes: the bell is there, the assistant comes, he rings the bell, the dog salivates, the meat is produced, and so forth. And then comes the moment, the Pavlovian *experimentium crucis*: The assistant should once more ring the bell but not give the dog any meat. Only this time, without the assistant's knowledge, Konorski removed the clapper from the bell, so that the assistant comes in, shakes the bell, everything stays silent—and nevertheless the dog clearly salivates. "Aha," we can say, "the ringing of the bell was a key stimulus for Pavlov, but not for the dog!" Here you have a further variation on my claim that the listener and not the speaker determines the meaning of a sentence. What I find very sad, however, is that Pavlov received the Nobel Prize for his key stimulation—but the more profound Konorski went away empty-handed.

If we take this conceptual pairing, perception and change in movement, then we come to an interesting interconnection between sensorium and motorium.

Ah yes, now you're pointing in another direction, in which the idea of movement and the appearance of movement become important. And here is the fundamental question about this: Does one perceive simply by being a receptor? At this point, Popper would use the picture of a bucket.[19] Earlier experimental psychologists believed that for perception one needed only a layer of photo-sensors—something that is sufficiently photosensitive—that had behind it a complicated apparatus that would work through these retinal pictures in the manner of a photo lab. For a long time the discussion ran in the direction of depiction and photo editing. But against this background, serious problems soon emerged. For example: How can we experience spatial depth? We haven't got any depth organs. At first people fell for the excuse, "We have two eyes spaced a slight distance apart, which necessarily see two different pictures—and from the difference between these two pictures we can reconstruct spatial 'depth.'" Initial serious doubts about this explanation were soon raised, for example by Poincaré.

Henri Poincaré was interested, amongst other things, in space-perception and played very intensively with the problem of multidimensionality and the possibility of its spatial perception.[20] He thus asked his psychologist colleagues, "Tell me, how can I see properly at all? After all I have a retina that is two-dimensional. How can I recognize an area in space?" And his colleagues answered, "My dear Poincaré, you have two such retinas, which both see different surroundings, and the combination of the two pictures forces you to postulate depth." Poincaré, a mathematician, sat down and asked himself, "Can I represent this process mathematically, can I calculate these depths? I can only do that if I assume that these two pictures are pictures of the same objects. That means I have to make an assumption about the identity of objects in my field of vision. But how should I know which objects are the same? Who tells me, who helps me along?" In all these lovely scientific tales there is quite a good dose of mythology, so we might imagine our case as follows: Poincaré began to shaking his head over how anyone could fall for such a contradictory theory. And as he moved his head rapidly back and forth, he suddenly saw that his pictures were immediately changing: "Aha, I have to change my perception, then I can conclude from the change in perception of what I've just seen because then I'm sure that I'm seeing the same thing, just from another angle!"

Poincaré introduced this new idea, namely the change in perception through body movement and with it created a "depth sense" from motoric movements. If I change the surface of a body, the consequences of this changing for the surface of my body and the change in perceptions connected with it allow me to reach conclusions about the three-dimensional arrangements of my surroundings. This insight, that the motorium, as I call it, feeds back into the sensorium and that the interpretation of the sensorium by the motorium interprets the movement of the motorium, leads to a cycle of mutual interpretation of the activity of the sensorium and the motorium—and to a unified representation of perception. We achieve the integration of perception through the motorium and the unification of movement through perception—and the importance of this double coupling is brought to bear most beautifully by Piaget. In this sense movement is a prerequisite for perception—and perception a prerequisite for movement. The two are so intensively coupled to one another that if you cut the two cycles only blindness and stiffness remain.

But at another point you emphasize the asymmetry between the two when you write: Every change in movement means a change in perception . . .

Yes.

. . . but not every change in the sensorium means a change in movement.

Right, yes. What does that mean? That means that sitting here you can think about all manner of things, a Schubert song, a Raimund couplet, and so on. These are actual perceptions, of course—you sit there and listen to the *Forellenquintet*, the *Hobellied*, you listen to that. Naturally from time to time you don't know how it goes, but you invent transitions, jump back to familiar parts—but no motoric change is needed for this. You could of course, if you wanted to extend or generalize it, say, "In the cells of your body there is always enormous activity." You cannot play the *Forellenquintet* internally to yourself unless you move impulses back and forth in the auditory segment of your brain, so that actually *"In einem Bächlein helle . . ."* begins with you, in you. Movement, though internal rather than external, is constantly going. In this case the idea of asymmetry is invalid: I no longer need external movements to be constantly changing. The body's internal states prove sufficient in this instance.

Our simple machines in our game box—not only can they move, but they also possess the capacity for reproduction.[21]

Yes, yes, quite. You're asking me how the capacity for reproduction came into the world? And my answer: No idea, I don't know why; it just happened that way. In most cases we really don't need to ask why at all. Anyway, I don't want to do like the others and explain an unexplainable thing in pretty poetry—or perhaps prose? I'll leave that to the evolutionaries.

That brings me to an important point: What is evolution? For me that is a central question. For me diversity and death represent the two fundamental principles of evolution, and not different degrees of selection or fitness. In my opinion, that's even a misleading claim: Why "fittest?" This ensemble is here by chance; why should it possess a greater "fitness" than another, why is one "fit" and the other not, why are these "fitter"? My basic question is thus, "How does this tremendous diversity come about?" The greater the diversity, the greater the possibility of adapting to a practically endless variety of possibilities. Conversely, death is an important process through which systems, ensembles, fall back out of this diversity. Thus, diversity and death stand together at the center of my idea of evolution.

Death of the species or death of the individual?

If all the individuals disappear, then the species has also disappeared, although I would generally warn against speaking of "species." The species is already an abstraction. I would say, "Death of the individuals, diversity of possibilities."

What matters, however, is not that a single dinosaur dies but that all the dinosaurs leave the game box.

The dinosaurs are still different as individuals, however. If Mr. Dinosaur dies—and the rest along with him—then the whole species has also died. The species is a consequence of our distinctions; we call it a species. Dinosaurs have disappeared because all the dinosaurs have disappeared. I don't see it this way: "The dinosaurs as dinosaurs have disappeared, therefore they are all dead." I say it just the other way round: "They have all died, and therefore the dinosaurs are gone." Some would like to say that the idea of dinosaurs no longer fits into a changed environment, and therefore the species died. I prefer to consider it the other way round.

With diversity and death we have one possibility of description. What's conspicuous here is the absence of the concept of selection, which usually plays a central role in evolutionary formulations. Is selection for you an unsuitable construction, a false, a misleading metaphor?

Nothing is chosen; things just die. If someone says to me that nature is selective, I reply: That's similar to the paradigm shift. Nature chooses nothing. There isn't a great Selector sitting there with his tweezers and his tongs, meticulously removing the dinosaurs from the playing field; they just disappeared. It's exactly the same, by the way, with the strange terminology of attractors—nothing is magically pulling. And in the case of evolution as well, nobody chooses, nothing and no one selects. Why then are there nevertheless these clear differences? An elephant is different from a camel or a rhinoceros. Here the basic idea already proves false and unsuitable. There is no selective apparatus that certifies the rhinoceros with one horn but rejects the relatives with two, five, or six horns and selectively banishes them into the void. Here we have to fight against misleading concepts. If I maintain these confusions, I immediately start to think about resulting problems, the bases of which, however, are already warped.

If I'm sailing under a conceptual flag that, like selection, forces me to think in a certain way, which proves not only not very constructive but actually barren, then I would like to ignore such a leitmotif completely and see whether I can find new leitmotifs. Here I would rather follow Ernst von Glasersfeld, who has developed a really lovely idea of evolution in which he introduced the word "fitting." 'Fitting' not in the sense of fitting in, but fitting like a key in a lock. There are those keys that fit and those that do not fit. If a fit is found, then I can open the lock and go through the door. Ernst uses this metaphor as if to say, "I don't know how the lock works—I just know how I get through the lock." I really like this idea of fitting; it isn't selective, rather it allows and forbids: "Now I can't get in there."[22]

The Maturana interpretation of evolution is very similar.[23] Maturana introduces a new concept called "natural drift." This drift idea assumes that internal changes are happening constantly, and these internal changes create new diversities out of which more possible fits arise—it doesn't matter who or what is supposed to fit here. This "fittingness" also requires an external world and makes this or that supposition. These demands are completely harmless in the course of conversation, but if one declares them to be fundamental ideas for what becomes the case—and I stress "becomes"—then I would say, "Listen, please be extremely careful!"

This idea of fitting does of course seem essentially more beautiful than the idea of selection, which one can so easily criticize. Even the sixteenth-century Calvinists said, "We are chosen because we are here." Much in Darwinism functions in completely similar ways. We are here, we're sufficiently fit because we—since we're still living—have passed the selection tests.

I'm really delighted by this example.

Only, the problem of the Glaserfeld keys could be that in the end this idea comes down to the same processes of selection and choice. I have four, five keys with me; one or two of them open the lock on many occasions—and therefore I'm going to take the others out or put them away.

Even as a child I was unhappy that we used these fighting-metaphors of "survival" and "adaptation." Why survive? Why not live? Am I just about surviving because I'm amusing myself talking with you? There is life, there is also fun. In "survival" I would constantly be facing death. The baroque idea that it's about survival when right now I'm alive is actually an embarrassing thought. And then "fit" and "adapted"! Why am I fit or fitter than I was five minutes ago; through what exertions or contortions have I become better adapted to my environment in the last few hours so that I've been able to survive since getting up this morning? This whole language is alien and repulsive to me, and therefore I'm immediately looking for all possible conceivable formulations besides selection and survival. The conceptual pair diversity and death, that's incomparably more congenial to me. There's nothing about fitness there; there's no fighting and struggling for survival; there's only diversity, which surrounds us, and death, whereby this or that perishes. Otherwise expressed, elements are not infinite but finite.

Death doesn't select either; we just die.

That's it. And naturally with his idea of "drift" Maturana is referring primarily to molecular biology. What happens here? A genetic structure is postulated, an area is developed, voilà, and in it new games, new moves and strategies may be tried. In a certain sense I like the Maturana game "Evolutionary Drift" very well because one needn't provide a detailed breakdown of mutation and selection processes. Instead, one can generally conclude, "A nontrivial machine can change its operations internally so that its drift results in an extension of diversity."

What is going to give us all a headache is the concept of "sort" or "species." In the fields of botany and zoology, from the very early stages, a lot of energy has been invested into the designing of taxonomies, into finding certain types of plants and animals or specific "kingdoms," into distinguishing and describing. Were these taxonomical efforts sensible, or were they just distinction therapies?

Taxonomies give me insights into the thinking of the taxonomists, the relevance of which varies with the originality of the taxonomy. The great taxonomy of taxonomies of Aldrovandi, for instance, gives me great pleasure. In, I think, 1600, Aldrovandi wrote two or three volumes of taxonomy about insects, mammals, vertebrates, invertebrates, and so on.[24] This taxonomy is so beautiful because you can so clearly see the connections: animals that Aldrovandi loathed—let's say, cockroaches, worms, jellyfish—are all put into a category together. And if you look at these pages then you see the horrifying sights unified into taxonomical harmony. On the second page the noble animals come together, a horse, the giraffe, and the like; on the third page are the wild ones, there you find maybe a rhinoceros, that belongs to the wild category. One sees straight away that Aldrovandi thinks just like you or I: If I see a rhinoceros before me I run away; if I spy a cockroach I'll step on it; if a jellyfish comes along I want nothing to do with it—except at Duarte's Tavern.[25] It is a taxonomy that I understand very well and can comprehend and with which one can create wonderful drawings, fabulous animal compositions. If you deal more closely with such taxonomies, you'll come across many authors who reveal the taxonomist's way of thinking to you. "All animals that have four feet belong to the quadrupeds." "All animals with two feet belong to the class of bipeds." "All animals with a thousand feet belong to the millipedes." Or, "All animals with green hair and blue eyes are the bicolors." If their hooves are red as well, they're "tricolors," and so on. There are an immense number of categories to choose from. And among the taxonomists, eventually it was Linnaeus who used similarities as the important criteria for distinctions.[26] There is, for instance, a four-footed ur-frog at the beginning. For this four-footed frog there are found the following similar variations: similar, similar, similar, similar. Darwin was also familiar with this taxonomy, which was quite popular at the time. My feeling is that this situation must have been similar to the case of Kepler, who found that one can bring the planetary orbits into ellipses. Newton developed this arrangement further: "If my hypothesis of gravitation is correct, then this planetary ensemble must work like this"—and finally he found those laws that explain the planetary movements.

Likewise, Charles Darwin must have considered these taxonomies. "My God, they really do all display very great similarities. If I put these types in a row like this, then it looks as if one developed out of the other, that means, the end result is the result of a previous step." And suddenly the idea of evolution occurs to Darwin. Regarding the question of taxonomies, however, it means that Linnaeus's taxonomy contained within it a possibility of an explanation that the other taxonomies lacked. In just the same way, epicycles presented great problems to consistent explanation, whereas ellipses almost provoked an explanation.

Although any taxonomy at all is possible, under certain circumstances some taxonomies invite the observer—through the specific distributions of forms and figures—to various associations and finally to the creation of different theories. On the other hand it could of course be claimed that the Pythagorean bodies, tetrahedrons, cubes, and so on, developed out of a fundamental form, an ur-body, or were created through foldings and subtle folding processes. At any rate, I like to see whether A may be traced back to B because then through that I can better understand A and B. Here we are dealing with various cognitive abilities that allow me to bring forms into relations. But with this we have—as so often—landed in the area of language and semantics. Using semantics, can I get operators going or allow operators to emerge that will help me to create relations between various experiences, to create a relation between a frog and an elephant or between giraffes and cockroaches? If I manage to build up a relation by talking now about evolution, now about diversity, now about death, then I'm in a comparatively good position because then I'm in a position to categorize a lot from my environment and I can construct diverse possibilities of relations between elements of my experience.

We can conclude with Umberto Eco and Heinz von Foerster: Relations are always there, one just needs to make them—or with a little inversion: We begin with naked names—and bit by bit create colorful roses from the netting of relations.[27]

FOURTH DAY

Cognition, Perception, Memory, Symbols

Cognition → computing a reality.
—HEINZ VON FOERSTER, *Understanding Understanding*

The prophecy does *not* run, that a man will get *this* result when
he follows this rule in making a transformation—but that he will
get this result, when we *say* that he is following the rule.
—LUDWIG WITTGENSTEIN, *Remarks on the Foundation of
    Mathematics*

And God said, Let there bee lights in the firmament of the
heauen, to diuide the day from the night: and let them be for
signes and for seasons, and for dayes and yeeres. . . . And the
euening and the morning were the fourth day.
—GENESIS 1:14–19

*In our fourth conversation, new possibilities should arise for us, of distinctions,
possibilities of orientation, of remembering—and of forgetting. So far we have
been operating with something that we could call an "MS system." In our game
box we found recursively coupled motor-sensory systems, MS systems. A further type in the box is the MBS system—a brain interpolates itself between the
motorium and the sensorium . . .*

Now I see where the B comes in.

The first type has a single recursion. The second is repeatedly closed recursively. Can we pass review on the important differences in the architecture of these two systems?

Of course I could say a lot about that—but I don't want to. I would much rather get back on my hobbyhorse and invite you to gallop away with me. Our theme is cognition. I want to point out that one cannot talk about cognition without cognition. The problem for us, therefore, is not so much to describe types of toys but in being able to describe toys while we ourselves are toys in this toy box. This problem is hardly ever taken up, because it isn't seen, or because if it is seen, it is suppressed and pushed aside. One of the most common excuses is: If we talk about ourselves, paradoxes arise. And in a logical system as important as biology or psychology one may not open the door to paradoxes. For me, this argument is a bad excuse. Another reason why people don't like to touch on this problem further, even if they've seen it, is that it requires a whole new form of problem solving.

We are so trained in always solving the problems of others that we hardly have any time left for our own, let alone for solving them, that's why problem-solving therapies in nonpersonal matters or outsider-psychotherapies are flourishing. Interestingly, the circular dance continues because therapists, too, are primarily concerned with solving the problems of others, hardly ever with the solution of their own. The solving of one's own problems requires its own solution: a certain form for posing the question and a form for creating something within this question, which one can consider as a form of solution, alone or with others.

For me, therefore, it poses much less of a problem to say, "Ah, a brain emerges and develops, a wonderful apparatus with so and so many buttons, which can perform these or those operations. Put these operations together, and then if you input 'pop, pop,' 'peep peep' comes out." Such devices can be put together in manifold ways, and in this area there certainly are witty solutions, monotonous architectures, baroque blueprints, nonsense formulations. What interests me lies in another area: "How does a conversation structure itself if I'm earnestly dealing with a problem, the successful solution of which the conversation about the problem already assumes?" The problem of cognition must already, in an important sense, be solved in advance if I start talking about cognition. Everything that I explain now about cognition, motor and sensory functions, the brain and nerves, is already anticipated because we not only possess the aforementioned abilities but have also learned to work with them successfully.

It would be lovely if in the course of the conversation we could strive toward and achieve the goal of drawing attention to these fundamental problems. If we climb up, down, or into the world of mental phenomena, then our problem is, "How can we talk about mental or linguistic phenomena with the language and the mental capacities that we've taken as givens in these questions?" I repeat Wittgenstein: "What is a question?" "What is language?" If you ask, then you already know how language operates. How can you ask me? You know it, I know it, you can do it, I can do it. But if I can do something, I still don't know how I can do it. If I have before me a nontrivial system in equilibrium, with an eigenvalue, I can only say, "This eigenvalue has developed itself—who knows why."

It's true that by changing perspective in the manner you've suggested, one does come to another way of presenting the problem. I don't however see the new forms of problem solving yet. Possibly they are lying within it, but not yet developed.

That's exactly what I see as well: It is still a research proposal. And I would even say, "If you use the world perspective, then it's about the problem of perspective, about the perspective of perspective—the perspective puts itself in perspective, as it were." Because my hobbyhorse, the eigenbehaviors of closed systems, the operations of which we cannot, however, identify, also gets into the game. We can only identify that there are eigenbehaviors and stabilities, that I can say the word "perspective" and others know what I'm talking about. I also see our conversation as a kind of research-support program, much as you, Albert, have also seen it. You said, "How could we develop a perspective, etc.?" How could we? The problem is not yet solved. It is often claimed that if one can name a problem then one has actually already found a solution, that is, one has shown in which direction one must search to cope with the problem successfully. We could also claim, "You have pointed out the shortcomings in the usual approaches to the problems with which we are normally concerned. Pointing out these shortcomings can prove quite helpful." Without the support program we spoke of, this rather unsatisfying but normal situation arises, which, speaking abstractly, is as follows: "We already know the solution, it is such and such, I wrote this book on it, I received the Nobel Prize for it, and so on and so forth." However, if one pulls back the curtain, one suddenly sees a gigantic black hole of ignorance.

In Olaf Breidbach's interesting book on brain research in the nineteenth and twentieth centuries, one could point to many examples of the following "accident logic": One starts with a normal, competent brain, and whenever dramatic accidents occur, then the doctors assemble and begin to examine brain and patient to see what defects arise.[1] Then from the absence of a part of the brain that has been destroyed is concluded the absence of an ability that the patient had before. The inverse operation functions the same way: If a person considered a patient couldn't display a certain behavior, then the accident-logicians said, "A piece is missing in this corner, let's stimulate the surroundings with electricity!" What kind of logic is that, treating it that way? Interestingly, we "localize" our more complex trivial machines according to the same pattern. If a power distribution network breaks down, then, as in Ashby's anecdote, a defective light bulb, a porous wire, or a faulty circuit soon turns up as the main cause for the power loss—and the distribution network and its structures have completely disappeared from this method of observation.

It seems that the logic in force is the kind very aptly characterized by Eilhard von Domarus, a young and very brilliant friend of Warren McCulloch's: "Someone loses their left eye, now they no longer have stereoscopic vision, their depth perception is lost. First reversal-conclusion: The possibility of depth perception must be located in the left eye. Now it is shown, however, that depth perception is also lost if one loses the right eye. Reinforcement of the reversal-conclusion: So the left eye must possess a suppressor that suppresses depth perception should the right eye be lost."

One can spin this false logic further and further, the whole ball of wool tangles itself into grotesque nonsense. It starts with the observation of a defect, one then ascertains the absence of a piece of tissue—and explains the missing part as the generator for the operation that, because of the tissue failure, can no longer be carried out. We encounter the same illogic or defective logic in the case of memory. It starts with the seductive language that invites us to separate functions that, operatively, cannot be separated at all. We speak of "memory"—and already one thinks, "There must be a warehouse in which all the contents of memory are stored in chronological or alphabetical order." Or we read about "depth psychology," ah, under the surface there must be a "secret safe" in the depths that would reveal all our secrets if only we drill deep enough and are able to open it. Language invites us, in a very charming, seductive manner, to draw distinctions that, in conversation, are understood very well—"He doesn't look deep enough," "This person has a good memory," "It seems that he's lost his memory!" But it proves nonsensical and misleading to identify specific containers or areas of the brain that create exactly these functions.

The localization of memory functions is, however, a very common game. It would probably be impossible to get you to collaborate on a brain atlas?

I very much hope so.

On the other hand, if one goes into a neurology clinic, there will be many people who actually have had an accident, who actually have lost a piece of tissue—and can no longer walk on their left foot, can no longer move their right hand. How does this simple empirical knowledge, which every doctor is familiar with, relate to your nonlocal approach?

I reach for a simple image: If I'm wearing a knitted cardigan and I tear off the woolen thread at one point, the whole cardigan suddenly unravels. Now we could say: "At this point, where the tear is, lies the essence of the cardigan." And because it is localized there, the whole cardigan disappears if I destroy this piece. What has then fallen out of sight is the network of threads, the "cardigan net." In a cardigan net in which one thread is connected with another, this kind of unraveling can come from a single point; destruction could always take place from any point. That does not lead, however, to the reversal-conclusion that the point of the cut was the place where a certain function, which now no longer exists, sat and could be localized. The system doesn't work anymore because damages at one point can spread to the whole.

I would like to make an additional remark: Humberto Maturana once talked about experiments with a cat, in which one cuts through certain nerve threads in the optic nerve—and the cat was blind. Now one cuts through a second nerve thread—and the cat can see again. In one case the whole system is so disturbed that vision disappears, and in the other the disturbance is transferred to another function, leaving vision unimpeded. There are important insights to be gained here. It shows us that one must always consider a cognitive system as a whole and not in the pseudo-correlations of "accident logic."

That brings us right back to the cat and its massive parallel networks. Cat nets are then not primarily based on localized, but on scattered circuit diagrams?

I would say that the assumption of a fixed and localized circuit diagram is very easy—too easy for interpreting the present problematic. Strictly speaking, nothing is interpreted here; one says, "Ah, A is connected to B." The laziness that one allows oneself in order to maintain these relations becomes dangerous the moment I start to deal with holistic problems— sight, memory, learning. And if I start to take an interest in the question, "What are the forms in which we can say something about the brain?" then I must completely give up these one-to-one correlations of "tissue" and "function." Because they immediately lead me away from any more comprehensive understanding.

Warren McCulloch once wrote a splendid and well-known article, "Why the Mind is in the Head."[2] Thinkers with a stronger systemic interest, such as Maturana, Varela, and others whom I know very well— including Heinz von Foerster, whom I know less well—see it differently. Maturana once discussed the negation of McCulloch's proposition, "Why the Mind is Not in the Head." With it he wanted to show that cognition is not organized in such a brain- or head-oriented way. If one isolates the brain from the muscles, cuts through the spine, then one very easily gets into a problematic landscape, in which "outputs" and "inputs" dominate and beautiful recursions have once again vanished.

Important current cognition theorists—such as Damasio, Edelman, Gell-Mann, Hofstadter, Holland, Minsky, and many others may all be characterized by Daniel Dennett's motto, "We are almost all naturalists today."[3]

My mind—and my body—delight in such a statement.

This dynamic network perspective seems to me to be central as a holistic motivator, especially concerning the recursive integration of the brain, motor and sensory functions. It would have important consequences if we learned to consider the processes at the S-end, such as vision, as a process of recombination between brain activity, the senses, and motor functions.

I have here the results of an interesting experiment developed with us at the BCL by Humberto Maturana and his assistant Gabriela Uribe. It shows that colors are a type of calculation from the retina up.[4] In this experiment, if I move the colored transparencies a little, the context influences the color-perception of colored squares that have identical wavelengths. With the same frequencies, the sensation is calculated once in this form and once in that form—we "see" different colors. Here again we find an importance in the argument against "portrayals," because obviously this phenomenon can't have anything to do with portrayals. It has to do with our sensorium beginning with a certain stimulation on the surface of the body—and not ending there. And this is why I find this experiment important—it represents a kind of entree into the problem of cognition and the question of perception. If we could manage to leave the idea of portrayals to the photographers right from the start, then I'd invite the reader or partner in conversation to think, "Then what's actually happening if I don't portray, if the idea of portrayals no longer works? How must we then talk about cognition, about perception, about sensation, and so on?"—And with this we have catapulted ourselves into a situation in which we have to press forward to new concepts.

My suspicions go even further: As soon as I carry out these separations and localizations and anchor the idea of the portrayal, more and more apparently "objective" problems arise as a result . . .

. . . also an interesting conceptual triad: portrayal, photography, objective . . .

. . . and my own freedom to decide seems to slip away from me more and more. I picture a picture of the outside world, I can't help it, the portrayal within me is like the model out there. In this way more and more of the choices that are available to a seer, perceptionist or cogitant, are suppressed—and the whole problem of responsibility is avoided.

In Greek physiology, for example, seeing was an activity—the eye sent out a vision-ray that scanned objects, and therefore the Greeks could "see" a goat, a tree, the temple, the statue of Zeus, and many other things. At the time there was an interesting objection: "If this vision-ray sight is correct, then why in Heaven's name can we not see in the dark? The vision-rays can be sent out their journey just as well in the dark!" The solution was that we need the light so that the vision-rays could spread themselves out; the vision-rays need light as an extension medium. Much later the camera obscura appears as a model for the eye. Maturana owns a wonderful book about the brain and perception. There the portrayal idea is so beautifully heightened that it is followed through ad absurdum in the book's illustrations or portrayals themselves. He adopted one of the pictures for his book *The Tree of Knowledge*, namely, "Caesar's way of recognizing the world."[5]

This picture looks something like this: Caesar's head cut open in profile, with eye, nose, mouth, and so on. Further, we see that Caesar's eye looks at an eagle before him, *aguila* in Spanish. A picture of an inverted eagle falls on the eye. The inverted picture of the eagle travels to an organist who sits in the brain. The organist sits before a colossal organ, which makes the sound A if he presses the letter A. The Spanish organist thus writes *aguila*. The organ pipes produce A-G-U-I-L-A; this command sets off marching to the organ pipes in the larynx; and out front, Gaius Julius says "*aguila*." And with this the whole problem is explained—the retina inverts the eagle, the organist thinks and steers—and the mouth obediently pours out "*aguila*." This form of explanation requires the church mechanics of the seventeenth or eighteenth centuries; it requires the colossal organ, the organist, pipes. I need not repeat myself again, but I find the form-problem of an explanation central. This specific form of explanation came out of a certain time. The scientific approach that expresses itself in such mechanisms tells me very little about how the eye functions, how perception is successful. We merely read the calling card of a person who wanted the process of seeing to be treated as a continuation of the organ stool with other means in different surroundings.

I also see a form-problem in our conversation if you ask me, "How do 'sensorium-brain-motorium' develop?" "What distinguishes 'SBM systems' from 'SM ensembles'?" and more of the same. With this you set a form of conversation and form of answer within which I should move and stay. The thing is just that I don't really like this form-landscape, but fair enough—I'll gladly take a look at this form and talk about this form. I would constantly like to stress, however, that SBM or similar variants represent a form that forces me to talk about problems of cognition that I don't actually want to see analyzed.

A side remark: The way in which the pictures of Caesar and the eagle are created, technological achievements, the state of mechanics or standards of production and craft play an important role. In one case it's an adapted organ, in another case it's a miniaturized photo lab—one mostly makes use of known sociotechnological repertoires to get models for the process of cognition.

Exactly, and most of these analogies are excellent thought and character sensors for the analogy-user. I would like, however, to return once again to the triad of sensorium-brainorium-motorium. Well, yes, if one enjoys drawing this distinction, then of course there are interesting insights connected to it. And one of those refers to a peculiarity in the long history of evolution: The ur-sensomotoric cells of ur-sponges, hydras, or very early multicellular organisms are directly coupled with a contractive element. If you tickle a hydra with acid, it immediately draws itself in—like a trivial machine. A little chemistry—zap, zap—and it contracts and retreats from the area.

The picture of Caesar's organist is misleading in several ways: For example, it also implies that perception is processed sequentially—first the outside world is turned upside down on the retina, and then deep within the brain it is set upright again.

These reversals were described in a very amusing way by the Innsbruck experimental psychologist Ivo Kohler. He conducted experiments with inverting glasses, which you surely know. One puts the glasses on—and the world is upside down, down is up, up is down. I once put on such glasses—it's horrible, at first it's like you're paralyzed; you move, and everything you do is wrong at first. The correlation between my movement and my vision of my movements is reversed and blocks my actions. Kohler got his students to wear these glasses for months. They weren't allowed to take them off, they had to wear them, walk home with them, wash with them, brush their teeth, go to bed. If the students survived the first couple of days, then over the following period they became more and more independent. What's fascinating about the stories these young people told about this reversal-experiment is that after the first days, everything within their immediate reach began to reconfigure. If they were sitting at a desk, they could pick up a pen, write poems, draw up protocols for the reversal-experiment, and so on. As soon as they stood up, their upright near-world broke down, everything turned upside down again, they weren't sufficiently reorganized. It's incredible how a new understanding was slowly built within them. And slowly, slowly, they reached the point where they could ride a bike through Innsbruck, climb this hill, look down at the valley, just like normal—everything looked upright and straight, just as in the time without the glasses. One of the test subjects said that on the first day of snow, she looked out the window and saw the snow quietly falling upward. Once she went out and felt the snow, the snow started falling from above again. The correlation between motor functions and sensory functions could not be better represented than through Kohler's experiments. One could, however, make fun of the simple mechanism of reversal. But it never is reversed—it is always as it is.

But the brain came into being because, among other things, the routes between sensorium and motorium—using the metaphor of telephone systems—bit by bit built up a control center.

This intervention of a B in your MS-world, it appeared very early on in evolution. I already explained about the elementary multicellular ensembles that functioned like trivial machines. Our descriptions of cause and effect work best with them, if at all; with them, one could say that cause and effect are localizable. But bit by bit the motor and the sensory elements became distanced from each other, they grew apart from each other in space. At first it was always the same channels or neurons that established the transitions between motor and sensory functions. But over time the distance became too great, so an intermediate element stepped in. In physiology this is called internuntius or internuntii, intermediate messenger and messengers, respectively. And this intermediary suddenly transformed this trivial system or this ensemble that could be described as trivial into a completely and totally nontrivial system. And it's probably for this reason that the internuntius built itself up incredibly quickly in evolution. The internuntius was, if you like, the first calculator in the long history of our world. With it, a yes or no, a calculation, a working out suddenly becomes possible—and the whole system becomes nontrivial. The moment a successful cascade or dynamic between senses-internuntius-motor functions develops, a system becomes no longer explicable from the outside. That is, unless science manages to so reduce the complexity of the internuntius so that the entire ensemble becomes a trivial system again and regresses. Naturally, that can be done, one cuts its strings, one takes away the connections, one dopes it with strychnine, and so forth. The result then works wonderfully: The system has become trivial—and I can publish all kinds of clever things about the by now broken system.

In what follows, we want neither to take apart nor to dissect these nontrivial systems, but instead to observe and follow their further developments through evolutionary history. An extremely important point in this respect is that these ensembles have become involved in their games with the world in a surprising manner, namely, in that they continue to code the outside world nonspecifically and show little specialization on the surface. "More means different?"

In the nineteenth century the German physiologist Johannes Müller had already made the following observation: If you take some sensor, the gustatory papilla on the tongue, for example, and you stimulate it with a drop of acid, then you say, "Ah, I taste something sour." "Excellent, very good, the papilla works." Now you use a drop of water, and as a reaction you get, "Sour." "Yum." Now you take a little probe and you stimulate the sensor with a little electrical impulse. "Sour," answers the papilla. However you work with this sensor, the sensor replies, "sour," "sour." Here we are dealing with an undifferentiated sour-sensor. If you now go to any other sensor, like a pressure cell here on the skin, a subcutaneous pressure cell, then you press it, and it groans, "Pressure," "Stronger pressure," "Weaker pressure," and so on. Now you stimulate it with a little electrical impulse, and our cell says, "Pressure," you give it sulfuric acid, it answers with "pressure," "pressure," "pressure."

From this, Müller developed the theory of specific nerve energy, which you now find in every relevant textbook on page seven, nine or twelve—at the beginning at any rate. Every student has to be able to rattle that off if he or she doesn't want to be a student forever. But after they've learned this principle off nicely for their test they go back to their laboratories—and again start to portray the external world. There's a fascinating cycle of learning and forgetting behind this, an exceedingly interesting series of working models. For in the lab, "green," "yellow," "acid," "oil," and all the rest of the salad is once again portrayed, represented—you'd hardly believe it. Here is my challenge to the historians of science: How is it possible that this subject matter was "learned" for over 150 years, from generation to generation, and ignored by the same people in the course of their further researches? I believe that Humberto Maturana, myself, and a few others were the first who really scratched our heads and asked, "How is it possible that we can see a correlation between this and this even though nerves tell us nothing specific about the 'outside'?" Nerves merely code, "I'm at this and this part of the body—and I feel so and so much." That's all you get. As we were dealing with the Müller findings, the Wenner-Gren foundation organized a conference in Chicago on the theme of cognition. Maturana and I appeared there, and our two papers are complementary; you can bind them together, back to back, belly to belly.[6]

The following phenomenon now stands in the closest relation to undifferentiated coding in the history of evolution: If and because it is encoded in an undifferentiated manner, the overwhelming effort of decoding has to take place within such nontrivial systems. One could claim that behind every successful sensor of the external world there are approximately 100,000 dedicated internal sensors. For me this relation opens up a wide field of wonder and great astonishment.

Before we go farther down this line, let's take a look at a nerve cell. There we see that a cell, that is, its corpus, stretches its arms in two directions. In one direction there's a totally straight, totally smooth fiber called the axon. It only splits in two from time to time, a bifurcation, so to say, and it has only a very few ends. In the other direction many branches stretch out, like on a tree, which is called "arborization," branching out, ramification—and on these branches there are hundreds and hundreds of little sensors.

Now, these little sensors can receive electrical impulses from other nerve cells. If an electrical impulse hits a sensor, it can make the cell perform various operations. One possibility is to get the cell to pass on this impulse; or maybe the cell says to itself, "Careful, if impulses come from anywhere else, ignore them!" Such a reaction is called "inhibitory." In various parts of the brain one fines various types of such cells—like those that, for example, ramificate with hundreds and hundreds of arms, from which they receive signals from all directions, other cell types where this tree structure seems somewhat pruned. In the cortex, for example, from what I've learned, there aren't very many of these sensors on the dendrites or branches, but even here their magnitude reaches maybe into the hundreds. In one region, the branches grow in their hundreds of thousands—if you look at one of these cells an unbelievable picture emerges. In this area the greatest contributions have been made by researchers like John Eccles and, in earlier times, Ramón y Cajal, the leading "brain mind" at the turn of the nineteenth century.[7] Every one of these sensors is waiting for the chance to react—they're on the lookout for electrical impulses. And these electrical impulses, as I've already said, can expect different reactions from the sensors. If you count all these sensors that sit in the brain waiting for each other, you get numbers and magnitudes that are just hair-raising. If, as you've said, there are maybe 100 million sensors and sense-cells for the external world spread over the body, then this number is practically nothing compared to what is in the brain itself.

If the brain is seen as a sensorium, then this sensorium proves infinitely richer than the sensorium for the outside world. From outside we hear almost nothing and see almost nothing. From within we hear and see constantly. Essentially we are listening, not to music, but to our own brains, our own heads. Normally we're interested in the outer casing—ears, eyes, nose. But no: we should concentrate on the brain, which is almost infinitely richer and more diverse. You can imagine it like this: "Close your eyes! And then what are all the activities going on inside—bad conscience, good conscience, you are pleased, you are frightened, you imagine something, you don't imagine anything, an incomparable inner waxworks. The cerebral orchestra turns out to be a gigantic ensemble compared to the little brass band that we see or hear on the outside.

With these counterintuitive relations of inside and outside—one thinks of philosophy and its models in the area of perception and the senses—the inside-outside relations are also subtly shifted. In one of your articles we find really astonishing diagrams of what a cat hears from the outside.[8] In the diagram we see that it's only when the sound from without is coupled with food that the cat hears within. If it weren't so easy to misconstrue, we might say: No outer hearing without inner voices.

This experiment that you're referring to was sketched as follows at one of the Macy conferences.[9] A hearing sensation runs, as we all know, from the eardrum to the membrane, the basilar membrane, across which many sensors are spread, then on into different stages in the brain. During these stages the impulse, created by the first sensors, is processed, exactly as is the case in the postretinal network of the eye. Then comes what I call the working out or calculation of these impulses; events proceed to the next stage, from there to the next, and so on. Eventually, finally, everything leads to the auditory cortex.

The experimenters developed the following experimental arrangement: They inserted a "cat microphone" at all these points, a little electronic listener, so that they could measure the intensity of the neuroactivity at these points. At different points along the acoustic chain of information, they could observe how much and with what intensity the cat did or did not hear. Now the cat was placed in the following situation: She goes into a cage in which there is a little box, which can be opened with a lever—if the cat presses the lever with its paw, the lid opens. In this little box—we wanted to name it the Skinner box—there's a dead fish. If the cat presses the lever, the lid opens and she can eat the fish. And now sound comes into play as well: A sound machine produces a tone—peep, peep, peep—a peeping tone again and again, a tone is produced every second. And the connection between tone and Skinner box? The lid can only be opened if the tone is played; if no tone is played, the lid stays shut. This is a complicated arrangement for the cat until she has figured out some important connections.

The cat goes into the cage, smells the fish: "Wonderful, now I'm going to eat some fish." If you now look at the various measuring points to check the nerve impulses, at first you just get noise' that means there no associated actions and no correlations being made with the peep, peep, peep of the sound machine. The nervous system takes almost no notice of the tone. Now the cat slowly starts learning to make connections between the tone and the Skinner box, tests, tries, "Aha, *so*, I get my fish!"—and correlations of the activity at the individual stations and the cat's auditory canal start to appear already. Eventually, once the cat knows what is going on in the cage, she goes to the Skinner box, "peep" sounds, ping, she presses the lever with her paw, the lid opens—and the cat snaps up the fish. Now the tone is spread through all the auditory stations, "peep" is recognized at every stage—"Ah, now we can open the box." That wasn't the case before.

This reminds me, incidentally, of that very funny saying: "Seeing is believing." I, however, would reverse this saying—"Believing is seeing"—one only sees what one believes. If you know it's all about this or that, only then can you see. Seeing alone wouldn't be enough—first you have to believe what's going on, then you see. Incidentally, here we're dealing with a phenomenon important for sects, cults, and preachers' flocks—if someone manages to make people believe, then the poor souls are lost.

In German there's another proverb that suits our purpose, which comes from the bad old days in children's education: "Wer nicht hören will, muss fühlen" (Those who won't listen must feel). For our discussion, there is another possible meaning for the proverb, namely that emotions, feelings—desire, happiness, hunger, joy fear, desperation—also play a paramount role in what is heard or seen.

A completely decisive role, of course. And that's why it's so important in what mood one experiences something. "Believing is seeing." "Those who want to listen must feel." You can come up with endless examples of it. Uncle Heinz tells you enthusiastically, "You've absolutely got to see this movie, you'll laugh till you cry." You go to the movie theater, and after a quarter of an hour you say to yourself, "What a boring movie, what a waste of fifteen minutes." Everything that's going on here is simple: You went to the movie in a completely different mood from Uncle Heinz. Uncle Heinz was in an excellent mood when he saw the movie. Or: Uncle Heinz didn't expect much from the movie and was surprised at how funny it was. If you had been in a similar mood, then the movie would have been funny to you as well. Maybe you also set your expectations too high because the movie was described to you as particularly funny. This systemic connection between believing, feeling and perceiving directs our attention to the following problem as well: In what condition was this gigantic network first exposed to a sensation? I can easily imagine that first experiences, first adventures, first feelings exercise a really decisive influence over operations and perceptions.

So far we have been dedicating ourselves to the rather playful aspects of this problem. Nevertheless, this kind of folk psychology—or as we might say in German, Küchentheorien, kitchen theories, isn't entirely harmless. Sometimes it leads directly into the realm of exclusions, confinement, the dictatorship of normality.

A very important point, yes. I suggest we start with Franz Joseph Gall's models.[10] There are countless pictures and descriptions originating from him in which an exact function is assigned to every region of the brain. What's unpleasant about these beautiful pictures, with the scalp arranged around them like a frame, is that they are still around today, the will-o-the-wisps of neurology. "Hmm, this man can try ever so hard to wiggle his finger, but it won't move. Aha, he shows severe damage to this region of the brain. So, this must be where the generator for finger-wiggling is located."

This problem persists in the therapeutic field with frightening consequences. I have a book on the history of treatments that were seen as appropriate at a certain time—dreadful. Let's just think about the separation of the frontal lobe of the brain, which was practiced in America for so many long years. One simply took a person and surgically cut through the part of the brain right behind the forehead in order to eliminate or switch off certain types of behavior. In this way I could "heal" a murderer because he won't bump anyone else off once the frontal lobe has been separated. Nobody worried about what else was eliminated or switched off, however, because people knew that the frontal lobe was where the criminal disposition was localized. This and other appalling therapies were widespread and based on these dreadful pictures of localization and the situating of abilities. And such therapies were used because people did not see these pictures as metaphors but took them more seriously, much more seriously, using them as models for the brain itself.

Rich possibilities for misleading and unfruitful metaphors abound in another area central to cognitive theory, the memory, which right from the start has been dogged by intuitive images of archives and monastic libraries.

I'll try once again to underline this point very clearly. The folk psychology version of memory: "Today I heard something new—'Two times two is four,' 'Africa is south of Europe,' 'Louis XIV reigned longer than Queen Victoria.' Aha, there must be a memory-chest inside me where these new things are stored, just like in a card catalog. And as with a card catalog, I can take little pieces, my memories, out of the chest again and again." Well, I could describe my car this way too: I turn my key in the ignition—and suddenly the car remembers how to be driven. It starts to vibrate slightly, and it remembers how to keep the motor running. The key that turns itself probably asks, "Tell me, my dear car, how do you drive?" And the moment the ignition key turns, the penny drops, the car remembers—vroom—off it goes.

Everyone will say this description is mad, you can't treat the problem of memory like that. My answer to them: Absolutely, I already know that. But that's how *you* treat memory. *You* claim, "There's a chest or a photo album in which everything can be stored. And now if I happen to set eyes on this or that souvenir, then this or that will come out of the chest or album." To get rid of such folk psychology, a new form of thought is needed. Once again our problem of form appears. How can I talk about memory without pictures and metaphors of boxes, warehouses, silos, chests, and other memory places? How can we talk about memory without making it seem like an isolated ability? How can memory be described as an observable function of a great flow of activities that one could call "cognition" or even "life"?

What do I mean by this? If I don't want to peel out the localization, the functional separation, but rather the great cognitive functions, then I need to have a language that pushes my attention in other directions, toward other relations and connections. And this form-problem is especially grave with memory. How do I talk about memory while simultaneously keeping alive the operative wealth of cognitive processes? If I were to tell the story of Hansel and Gretel, then you would say, "Aha, he remembers the fairy tale of Hansel and Gretel." But if I tell you a story about the trip I took last week, the word "remember" doesn't come into it at all. There I'm "telling" about a trip that happened recently. Folk psychology draws a neat distinction: "He says what he did yesterday," "He tells about today's visit to Duarte's," and not, "He remembers what he did yesterday," "He remembers today's lunch at Duarte's." Today we ate lunch at Duarte's, I "know" that—that I don't need to, or perhaps cannot even remember it. Apparently the telling is not at all connected to remembering.

The disappearance of remembering from the ability to tell is a problem that interests me greatly. What is happening for the other person, for my conversational partner, so that the idea of memory sudden vanishes before our very eyes? And then why do I need memory for "Hansel and Gretel"? Let's imagine that I don't need memory for "Hansel and Gretel" either, instead I constantly reinvent this story. Or: If the Hansel-and-Gretel problem arises, I choose those grunts and sibilants that will lead another person to think, "Here comes 'Hansel and Gretel.' " Let's assume that this story gets told differently every time, unless the Grimms' stream of grunts and sibilants has become such a habit for you that you are able to exactly reproduce the Grimms' version of Hansel and Gretel here and now. But where does this ability lie? I'll try to get another perspective: You yourself become Hansel and Gretel when you tell "Hansel and Gretel." That means that you—your entirety, your whole being, you as a person—don't just chew the meat, don't just drink the wine, don't just say to Annie, "The breakfast was excellent."[11] You're also capable of "Hanseling and Greteling." I'm trying to choose my language so that the problems of forgetting, of remembering no longer appears as a phenomenon of forgetting and remembering; instead, the linguistic shackles, which force us to think in terms of these separations, are loosened and kicked off. I haven't yet been entirely successful on this point.

This alternative idea of remembering would seem to encounter another massive stumbling block because of the storage model of memory used in technology, in computers. Memory and remembering are connected to time, and to a certain necessary length of time. Memory, remembering, these create a kind of distinction between people's presents and their pasts in social, everyday life. And the cognitive process of remembering is perhaps better described with another word, namely, the German word vergegenwärtigen *(to make present, to bring to mind). You would probably want to use* vergegenwärtigen *rather than "remembering."*

I'm really delighted by the word *vergegenwärtigen*. I really like it. There's no remembering a passive past, instead there's an active happening in the here and now. Hansel and Gretel develop here and now. I let an event emerge here and now so that the past is built anew here and now. I find that to be the beauty in *Vergegenwärtigung*; I'd like to include it in my vocabulary, with reference to you, Albert, who pointed it out to me.

Yet if I think about your book on memory from 1948, then there, without a doubt, memory is something that is a function of time and changes with it.[12] *Doesn't this old work stand in massive contrast to the conception that emphasizes the here and now so strongly?*

Correct, absolutely correct. But the 1948 memory book emerged out of a very specific problematic that arose in connection with my experiment: "Can I apply Ebbinghaus's insights to another question, that of whether memory or the memory storehouse can't be described in terms of quantum mechanics?" And it turned out that I had found a model that explained the data in a way that was totally new at the time. The question remains, however: "Can this earlier interpretation of mine be applied to memory-function in my current view?" Here, too, I think that something should be possible. I just don't believe, as Schrödinger himself didn't believe at the time, that molecules' various quantum states can be made responsible for it. Today I would stress more heavily the structures of relations in the brain that constantly activate themselves—and naturally these don't die off or regenerate themselves again. But I could apply some principles, which I developed for memory functions at the time, to such structures of relations and get similar results very easily.

We could briefly summarize your metaphor of the remembering car like this: What really matters is that you can operate the car here and now, can drive it, know its setup, know where the spark plugs are—the metaphor of remembering just disappears on its own.

I'd just like to stretch ad absurdum the idea of describing something that functions in a certain way as remembering. Surely we could also say that the car remembers that it is obliged to turn its motor over if the key is turned. Those are all lovely metaphors. The question is just, "Can I use it to fix the car? Can I get my driving license with it? Can I change the spark plugs with it?" The metaphor of the car remembering its duty can of course be applied to any problem. And this metaphor gives me an immediate understanding. "Ah, now I understand you, my dear car. Today you'll drive—I know you remember your duty to get me from A to B.

And during this drive, you, car, will remember how to turn right, how to turn left, and so on." "Oh no, today my poor car doesn't want to start. It has forgotten its duty to get me to Pescadero today. Please, dear car, just remember!" In principle, the words "remember," "remembering," and "memory" can be used for everything if seen this way. And this is precisely what happens in some of the humanities.

Let's forget about remembering for a little while. We have these wondrous networks of sensory-cognitive-motor functions that, day after day, are generated in the most diverse individual versions. What's interesting about this is that it is all produced from a genetic program whose informational content is incomparably smaller than the wondrous network that arises out of it.

I used to work with a very lovely and intelligent person at the University of Illinois, not professionally but on a friendly basis, it was my friend Henry Quastler. Quastler was a Viennese physician who fled to America after the Anschluss in March 1938 and worked in the Carle Clinic. He was an exceptionally conscientious, ethically and morally conscious human being, for whom the atom bomb represented a horrifying human catastrophe: "Can I now, as a working person, find out what damage has been done by the radiation of atomic bombs?"—that was his research question. Thus he started to conduct experiments on radiation damage in living organisms. For this purpose he bought very cheap little alpha-radiators and with them performed experiments on frogs and mice. His lab was tiny—a single room in the Carle Clinic in Urbana. I got to know Quastler by chance right at the beginning, in the first days after we'd come to Urbana.

Quastler very soon realized that the problematic he was dealing with was related to one's ability to qualitatively describe the damage caused by radiation. At this point he stumbled across so-called information theory and read Shannon and Weaver intensively.[13] He said to himself, "I think that this form of damage-description is very impressive; with it I should be able to analyze this topic." And he began, very early on, to convene conferences on topics like "Information Theory and Biology." And he was always calling me: "Heinz, you know Shannon, you're a mathematician, come, tell me: What is 'redundancy'?" I would answer, "Okay, Henry, let's sit down together and write the formula on the blackboard."

Henry Quastler learned the basic concepts and formalisms of information theory with a speed that was almost unbelievable. And why? Because he needed this instrument urgently. "A godsend, these formulas, splendid! I can go on now." And he'd run off again and get his experiments and his formalizations going. Soon after, he sat down with a second person, called Dancoff. The two recognized that you get interesting quantitative results if you use information theory concepts like redundancy, distribution of information, noise, disturbance, and so on. Then what happens if I use such information theory concepts in the fields of genetic programs and organisms? Thus the two of them, Quastler and Dancoff, developed a very interesting model, that made a lot of other people very happy as well, for the following question: "What is the information content of a gene?"[14] How many bits have to go in there? And what is the informational content of that which produces these bits? What—in the language of information theory—is the relationship between the quantity of diversity or complexity that this system can create and give rise to, and the quantity of diversity or complexity with which it itself has been built?

This problematic was repeatedly taken up by interesting people. Ross Ashby, for example, formulated this question in another form: Can a mechanical chess player outplay his designer?[15] Can an apparatus that was built with a limited wealth of information outplay the designer who designed the apparatus? Can it surpass him? And back then Quastler answered this question in the affirmative: "Of course, we can see it. If one calculates the proportions for a genetic program and then for a brain—essentially one just needs to count the synapses—then one hits on some really astonishing relations: The program that presumably defines the structure of the brain, comes up short of the required magnitude by a factor of about 10^{10}. That means that the genetic code, which determines much more than just the nervous system, is not capable of programming a network of the magnitude of the brain." Expressed differently, an architect, our dear designer-God, can sketch a relatively simple ensemble that can develop in unbelievably complex ways after it is built.

A very important bridging principle in this context is the following command: "To create something complicated, produce something very simple, but produce many, many copies of it and—'More means different'—make a 'big machine' out of them!"

Yes, but this big machine, the brain, for example, doesn't consist simply of assemblages that are copied over and over again. The important point is the interlinking of such elements. A form of interlinking develops in which every subsequent connection opens up and develops new possible combinations. We don't have Copy A lying next to Copy B, but rather Component A together with Component B—and vice versa. Gordon Pask would call this a "rule of superadditive composition." That means that the rule of compositions covers not only A and B and C but also the combinations AB, AC, BC, ABC. The moment components are connected with each other in this way, an unbelievable number of new relational structures emerge. There can also be connections, of course, in which components compensate for and suppress one another, which Pask would call a "subadditive composition." Superadditive and subadditive compositions—all kinds of things arise from these. These observations can even be used to draw conclusions about organizational and management theory, because here too cooperative or competitive schema can be represented or theoretically analyzed through rules association. Why am I putting so much stress on this problem of networking? Because I see this as an absolutely fundamental problem for computers.

If we take the great computers as an example, then to this day almost all of them are Neumann machines. They work sequentially, first one step, then the next, and the next, and so on, even if with the new processors that are always coming out this happens unbelievably fast and then even faster. Very early on, when I and John von Neumann got to know each other, 1949 or 1950, we got to like each other very much by talking about the possibility of parallel machines, in which the emphasis is not on sequence but configuration. We at BCL were the first to build, by means of example, a counter that doesn't count but rather examines the configurations, looks at them, and says, oh, "365." John von Neumann had a lot of fun with these considerations.

We've argued that sequentiality is not a necessity in calculations. You can, if you see a configuration, get it in one step. Do you have to solve a differential equation at a walking speed? Take the field equation of two magnets in space. First we'll write out this field equation—and then we'll calculate it step by step. The magnetic field is here the whole time, however; you can look at it and say, "Ah, that's a field with two magnets." The sequentiality isn't necessary because the unity of configurations can also be taken into account. I summarized our insights from that time in an article.[16]

This article probably appeared fifteen or twenty years too early to have the proper impact.[17] What are the greatest differences between the BCL machine and the Neumann architectures?

The difference was fundamental. In the BCL architecture, the relations-structure was built into the machine—it examined the neighborhood, the relations between neighbors—and figured out all the places where there were neighbors. It calculated the whole of an object through the calculation of neighborhoods. And now that's a parallel operation because it knows every neighbor to be a neighbor of a neighbor. Thus, at a single stroke, however many neighborly units come to light—and there was no need to count how many.

A parallel counting network, that is to say, a machine with a relational counting sense?

Exactly. The structure of *n*-ness must be built in, so that if you say "12" to the machine, it says "12" back.

A sequential machine that is ascertaining a numerical quantity or the number of a certain group counts sequentially, 1 and 1 and 1 and 1 after the other. How did your parallel machine do it? How did it get the numbers 377 or 124?

The parallel machine should just "see" how many objects there are. It should look at objects, not count them. If there are seven things on the table, there are seven objects to see—there is a seven-ness that manifests itself. A machine looks at it and says, "I recognized a seven-ness." Or take the six on a dice. I could now put three books here and three books there. If I arranged them like the pattern on dice, you would say, "Six." The machine does just the same. In the case of parallels, it's not about entering the number 6 and adding on the number 7; rather, it's about counting as the picturing of *n*-units in a symbolic form.

After this great parallel campaign and after our networking tours through the fields of cognition, let's move on to a possible endpoint, namely to the question of mechanical reproduction—and the possibility of automation.

An automaton is a very interesting thing, introduced by Aristotle. The expression combines the two words *autos* and *matizein*. *Matizein* actually means "to decide"; thus, "automaton" means "something that decides for itself." Aristotle described this point beautifully in *De motu animalia*. At fairs one very often sees traders who display little dolls that dance and jump about, move as if they had a will of their own. If one takes these dolls apart, however, it becomes clear that they don't actually have a will of their own. They're preprogrammed; through their construction they're made to move the way they move. They dance or jump like crazy because there's a little spring inside that drives the doll. Now comes the argument: "Suppose we couldn't take them apart, then wouldn't you think that they had a will of their own?"

One of the most famous automata in history was a chess automaton at the Viennese court that beat all opponents. Inside this automaton sat a very good chess player who actually won the games. The problem was to find a suitable chess player—and the trick went undetected for a long time.

There's an important addendum to that story. The original chess automaton was built at the end of the eighteenth century. Its designer also constructed all kinds of other automata, for example, the one that plays the trumpet in the Munich Museum. After the designer's death, the chess automaton came to America, where it caught the attention of a young man— Edgar Allan Poe. Edgar Allan Poe looked at the automaton again and again—to demonstrate that it was purely mechanical, one could open the front door see the mechanism inside. Poe describes this very precisely in an essay.[18] And Poe's conclusions? The visible mechanisms did not have sufficient diversity for such intelligent chess playing! In this observation, I would claim, lies the literary foundation of information theory. Ashby's "principle of requisite variety"[19] was not fulfilled, as we would say today.

What's fascinating about such thought machines is, among other things, that they can look back on a very long design history: from the Golem to hermeneutic traditions and the Renaissance to various conceptions in philosophy and literature—La Mettrie or E. T. A. Hoffmann, for example. In a sense, these automata are long-cherished dreams of mankind. What is the actual difference between the early pictures of thought machines and today's repertoire of artificialities—"artificial intelligence," "artificial life"?

I cannot find any difference here because neither case provides a satisfactory solution to the problem: A thought machine must be in a position to talk about its own automatism, to change it, and so on. The automaton, whether classical or contemporary, operates *as* something else. And this *"as something else"* represents a reduction of the necessary diversity. For the most part, our understanding of cognition and the models based on it are still working at the level of explanatory principles, a phrase that Gregory Bateson defined wonderfully in dialog with his daughter.[20] As usual, the daughter started with her questions: "Daddy, what's an instinct?" She immediately gets the answer, "Instinct, that's an explanatory principle." "But Daddy, what does this principle explain?" "Anything you want it to." "But Dad, don't be so silly. It won't explain gravity." Then Dad explains, "It's not usually used that way, no, but if you wanted, then instinct could also explain gravity. One could say that the moon has an instinct whose strength varies inversely with the square of the distance of the Earth . . ." "Dad, that's nonsense." "Yes, yes," he says, "I didn't ask about instinct, you asked about instinct." Memory, I would claim, is a Batesonian explanatory principle. Memory could also be used to explain gravity: "The moon has a power of remembering whose strength varies inversely as the square of the distance of the earth . . ." "But don't be so silly!" "Well, that's how most people think about memory."

We can also find such explanatory principles for other key cognitive concepts. For the area of knowledge or intelligence, one could use Henry Plotkin or his editors as an example, since the blurb for Plotkin's book Darwinian Machines and the Nature of Knowledge *contains the following text: "Henry Plotkin presents a new science of knowledge that traces an unbreakable link between instinct and our ability to know. Since our ability to know our world depends primarily on what we call intelligence, intelligence must be understood as an extension of instinct. The capacity for knowledge is rooted in our biology."[21] Here we find an everyday understanding thriving that furthermore promotes evolution as an explanatory principle. Instinct, intelligence, knowledge, biology—a typical word-rap that just keeps itself going.*

Splendid, simply splendid. How do you keep finding such gems?

And with regard to learning, the everyday assumption is, "A new little something extra goes into me—little pieces of learning for the memory vault."

On the one hand, yes—but you forget the Nürnberg funnel, the good old Nürnberg funnel! You drill a hole in the head, take a funnel, pour in the assorted letters and equations, and hope that the letters and equations end up in the right order.

The question is always: "What do these metaphors enable me to see?" What interests me is finding metaphors that don't allow me to evade my responsibilities. And if I manage that, then I run through all the various ramifications of this metaphor. I often see my friends, however, working with metaphors built to allow them to shirk their responsibilities. Then I pull on the emergency brake: "Can I get away from this picture? Can I sketch an alternative picture?" Then I say to my friends, "Hold on, it seems to me that this picture hasn't turned out so well—it seems irresponsible." I always steer toward the metaphors that allow me to express myself freely and that therefore make it clear that I am responsible for my answers, my activities, etc.

And our unavoidable responsibility to have some dinner is now becoming rather pressing.

I completely agree.

FIFTH DAY

Communicating, Talking, Thinking, Falling

> The interaction becomes communicative if, and only if, each of
> the two sees himself through the eyes of the other.
> —HEINZ VON FOERSTER, "Epistemology of
>   Communication"

> I want to regard man here as an animal; as a primitive being to
> which one grants instinct but not ratiocination. As a creature in a
> primitive state. Any logic good enough for a primitive means of
> communication needs no apology from us. Language did not
> emerge from some kind of ratiocination.
> —LUDWIG WITTGENSTEIN, *On Certainty*

> And God said, Let the waters bring foorth aboundantly the
> mouing creature that hath life, and foule that may flie aboue the
> earth in the open firmament of heauen. And God created great
> whales, and euery liuing creature that moueth, which the waters
> brought forth aboundantly after their kinde. . . . And the euening
> and the morning were the fift day.
> —GENESIS 1:20–23

On the program for today, we have the creation of the great sea creatures
and all living beings, which, according to your congenial British colleague

Gregory Bateson, can also be characterized by their ability to speak and communicate.[1]

For me, the first thing is that language distinguishes itself from a general idea of communication. Communication happens as soon as any creature waves some body part about so that another creature interprets this waving about, puts it into a definite relation with something and acts accordingly—for example, by getting "hopping mad." One can see this in mating rituals, combat rituals, and in the many other games of life that have been so exhaustively studied. From these communicative aspects and methods that one sees in animals—because they move these movements can be used to gain a "key" or "signs" from the environment—from this, I would like to distinguish language. Language is a special system of communication that can talk about itself.

That means that if one talks about bee language, then I would say that bees have—according to my definition—no language. Bees can certainly dance to one another: Buzz, the flower such-and-such is three hundred feet east–east–west—a fabulous medium of communication. But if the bee comes back, then it can't say to the waiting bee, "You said that very nicely, and it was all correct, but your pronunciation is catastrophic," or "Next time without the Texas accent," and so on. Bees can't do that—I don't think so, anyway. They also can't speak about their vocabulary, they can't talk about their grammar, their communication system itself is not communicable through their communication repertoire. Language begins, for me, when communication develops a concept of communication and becomes reflexive.

In another work, Bateson has also addressed the problem of communicative reflexivity. Bateson refers to the example of monkeys, who can signal to each other: "Listen, this isn't serious, just a game," or "Watch out, that's not funny anymore!" Through such rituals, Bateson claims, reflexivity enters communication.[2]

One can interpret it that way, but I would, however, say that Bateson's position is hard to maintain. That a certain interpretation is indicated through signs is not the same as talking about the signs themselves. If I use a certain gesture that says, "I'm not being serious now, I'm not going to beat you up," and if I use another gesture that shows that I am actually becoming aggressive, then my gestures aren't remarks on blows and aggression, but rather a modification of my next behavior, so maybe a behavioral reflection but not a communicative reflection—and thus also: not language. But my opinion is that the definition-question is unimportant—I'm not a definitions guy who runs around after definitions, and I don't want to quarrel about this matter at length. Distinctions like the one between communication and language serve me as guidelines, so that I only like to use the word "language" if the language talks about itself, if the language has the word "language," the word "word," the words "noun," "verb," "adjective," "grammar," and so on.

As soon as a communications process allows a reflection about the gestures that occur in such a communications process, then I'd say, "Now we're talking about language." If the communication's vocabulary has no word or no symbol for "communication" or "gesture," then according to my distinction I would not speak of language but of communication.

If I draw these distinctions, what advantage does it give me? For me, it's always an advantage if I can clearly distinguish between two areas and if I don't constantly slip from one into the other and from the other into the one. That means that if I'm interested in the language problem, as I see it, then recursivity and circularity will be moved into another area, namely into my own speech, my own life. In other cases we have circularity between two elements; that's called communicative circularity. If, however, I can say how I circulate, then I put this inside myself, then I have to deal with the organization of my nervous system or with the entire sensory-motor system before I myself can speak about language. That means that the structures I have to deal with will be shaped differently in the case of language than in a treatment of communication. But that's just pure Foerster methodology, which is what I use to juggle my ideas if I'm going to juggle them at all.

Thus, if anyone tells me about distinctions, then I always immediately try to clarify and operationalize the distinctions, so that I'll say, "If this category is set up, how many of these elements are there, how many of this kind of elements belong to it, and so forth." One might say I test the quality of a distinction through a quantitative analysis.

In the long-term development, communicative couplings had to develop first, and in humans, language formed in addition to communicative couplings—as communicative couplings of communicative couplings.

Yes, sure, of course, one could hardly explain this development otherwise.

But this recursive formalism is applicable—albeit with different usages—for both forms, communication and language.

In both cases recursions are set in motion. In one case I'm referring to the organism itself, in the other to the interactions of two organisms. Of course, the interaction of two organisms is hardly an exceptional event—it depends, however, on the epistemology that one wants to invent in order to explain the one out of the other. I have absolutely no idea how language developed out of communication. After all, there are people who claim, "In the beginning was the word and the man, the ape is a degenerated man, the crocodile represents a degenerated ape and the turtle a degenerated crocodile, they have 'forgotten' more and more—till eventually we get to the single-celled organisms from whom all communicative competency has vanished." This form of devolution also contains a wonderful process, although it's not upward but downward. I still know of—I can still "remember," it happened long enough ago—when I was young, there was a philosopher who claimed, "Man is the beginning of all things." There's a joke that fits very well with that, in which the son asks his dad, "Hey, Dad, did people descend from apes?" And the father answers, "You did, but I didn't."

Now back to your hypothesis, that one can only speak of language if one has a concept of language or if the language has a concept of itself . . .

I wouldn't want to call that a hypothesis but a definition. It's not a theory that I can prove or disprove. I'm just saying that if a communicative system can express itself about communication, then I would call this communicative system "language." If a communicative system does not contain or cannot contain this reflexivity about communication, then it remains simply a "language-incapable" communicative system.

The counterquestion comes at once, however: "Heinz, why are you carrying on with this ridiculous splitting of hairs, calling one 'language' and the other 'communication'?" Because I want to get out of the metaphor of language, so as not to carry the metaphor of language to bees or transfer it to jellyfish, because otherwise I'm dealing with the emotional life of jellyfish and am almost forced to claim, "This jellyfish really fancies a poem, but his wife doesn't allow that." It gets me into a question structure that is, in my opinion, tempting but that I don't enjoy. Maybe the perspective is charming to others, looking at poems about the love lives of jellyfish, how a jellyfish waves to another jellyfish, "I can't even tell you how much I love you." "See there, Heinz, there we have it again, this central element of language—'I can't even tell you . . . .'" And my answer to that? "Good, how lovely for you, very animal-loving and fair to jellyfish, but for me this nondistinction goes too far." It could be that other people also see an advantage in this differentiation. For me anyway there is certainly a clear advantage concerning logical structures and relations. And with it I can orientate myself.

Back again to your suggested definition: Language can be spoken of, and communication can be talked about as well—but communication does not allow itself to be communicated.

I would suggest and strongly recommend that language be conceived of as a second-order concept, namely, as a communicative system that can communicate about itself. For that reason the capacity for language also has a logical structure that has to be represented, realized, or manifested organically. If problems in their logical structures prove to be separable, then maybe I can think about how the manifestation of such logical structures occurs.

Perhaps we can work out some of these logical differences between forms of communication and language systems. What would be some further essential differences, apart from the principal difference, which is that one side can speak about itself and the other can't. What operations are available in the case of languages that don't appear in communicative systems?

Once one can talk about what one uses to communicate, one has the possibility of modifying how one says something. I can now suggest, "Let's use this gesture—tapping the forehead with a finger—to mean 'congratulations' and not to call someone a 'loon' and to say that they're stupid and simple." I can't put forward this kind of modification as long as gestures are explicitly loaded with singular meanings—and I can't communicate about the gestures themselves. If I'm now able to gesticulate about the gesture itself, if language emerges as a second-order phenomenon, new possibilities immediately open up: First of all I'm becoming clearer about the logical structure of what I'm doing because I'm able to reflect on my doing. As long as I cannot (and therefore also) don't want to reflect, as long as I can't willfully influence the modifications of my doings, they arise by themselves.

If a lion roars, "Wwhrrhauauuu," he's in a bad mood. If, on the other hand, one hears "Wwriieiihhnn," the lion's in a good mood. Thus far lions can understand each other. What communication of this kind cannot express, however, is roar sequences of the type, "Now I've been especially friendly, haven't I?" A lion can only give a cheerful "Wwriieiiihhnn." I could also say—and this assumption seems very justified to me: If a lion could talk, the other lions wouldn't understand it.[3] These sequences of roars and grunts could naturally change over the course of hundreds of years. But if they change, then they change and modify themselves principally because circumstances or environmental factors change, leading to such modulations.

But as soon as I can say, "Now that was a friendly grunt," or "Don't hiss so mean," then the sequences of grunts and hisses themselves can be talked about. Because I can talk about the grunts and hisses themselves, the sequences of grunts and hisses can be changed, modified or diversified into French, German, or Turkish. We get an enormous extension of richness in our capacity for relations. If Bateson's observation holds true—that in animal communication/gesture, relations are created—then this phenomenon is all the more so in that human, all too human, area, that is, the area of language.

If I can talk about language itself, that opens up the possibility for reflections on references and relations, on the weaving of linguistic relations, that means that I can modify references to references, and therein lies a very interesting possibility that multiplies the richness of possible references. I can for example say how much I want to be together with this one and how little I want to have to do with the other—and this not at all explicit, but rather implicit in the stream of my speech, in the stream of my dialogs. The importance of both cases, the case of communicative gestures and the case of linguistic expression, lies in interplay, however. They require each other to get this game going and to set it in motion.

With his concept of structural couplings,[4] I think that Maturana presented an amusing and wide-reaching idea. I'm not completely happy with the expression, but in this way the concept of autopoiesis[5] finds its natural extension. An organism A requires for the maintenance of its self-(re)production an organism B, which again needs organism A for the maintenance of its own self-(re)production. In this way structural couplings are established between A and B. Both mutually maintain their own autopoiesis. It's a very pretty logical idea. Now, Humberto can't really—and that means compellingly—establish autopoiesis out of this coupling, but this point is only of subordinate interest. As a metaphor, structural coupling proves to be an excellent idea. Take the lecture that Maturana gave in honor of the wonderful Eric Lenneberg. Lenneberg, by the way, was an outstanding linguist, an incredibly nice and intelligent young man who worked intensively with language and communication and who died young, very young, in his forties.[6] His wife organized a big meeting of linguists, to which Humberto was also invited.

Now, let's take another look at Humberto's article on the "Biology of Language," which in my opinion was one of the best lectures he ever gave.[7] In this work he brings autopoiesis and language together through the interplay, the structural coupling of two autopoietic units. If organism A requires organism B for the maintenance of its autopoiesis, then a structural coupling emerges, out of which can grow much communication between A and B, much waggling back and forth between two animals, but also "language": "Say, Dad, why do Frenchmen wave their hands about like that?"[8]

The engagement with language—talking about language—must have come to you almost spontaneously from the ambiance in Vienna at that time, with your parents' acquaintances, Uncle Ludwig, the Vienna Circle and Die Fackel, *the writings and lectures of Karl Kraus.[9]*

Surprisingly, my answer is no. For me the language problem, as we're working with it today, became its own area of work relatively late. Although I was enthralled with language and its magic from my earliest youth, language was still somehow a vehicle for me, a means of expression.

I first came to language as an analytic phenomenon during the work at BCL, where, among other things—out of the cognition research—we dealt with the problem of organizing libraries, translating texts mechanically, and the like.[10] At that time we dealt with the language problem as a problem of language—and one might say we had to learn and invent everything. As with many of the things that we learned at the time, right from the beginning we had the feeling, "That doesn't actually speak to the problem of language at all." A much-discussed problem in those days was, for example, artificial translation, machine translation. The first machine translations worked fabulously, as if out of a fairy tale; you'd have dictionaries for Language A and for Language B, and using an artificial editor turn a sentence in A piece by piece into language B. We got into this area more or less by chance and started to get interested in language. And that's when I reached the point of no longer regarding language as a vehicle but as a co-player whose inclusion can lead to new insights about cognitive functions, about the nervous system, and so on.

But your early reading of the Tractatus *simply must have moved language and the critique of language into the center very early on. The limits of your language were also—Proposition 5.6—the limits of your world?*

That's absolutely right. But with a proposition like "The limits of our language are the limits of our world," I take language as an already understood phenomenon—and not one to be investigated. The limits of the world are the limits of my language; there language has already been anchored to something known, language has already happened. The analytical language problem—"How do the limits of language emerge?"—that became a priority question only through the professional work at BCL.

It became clear to me at a very early stage that language is a tremendous magic charm—in that sense your allusions to the Viennese ambiance are justified. At the invitation of a world congress of social psychiatry, which had the motto "Farewell to Babylon," I wrote some introductory remarks on this theme. In this lecture I wanted to invite my listeners to not consider language in the usual sense of an explicable phenomenon—and thus I called my essay "The Magic of Language and the Language of Magic" and showed right at the start how magic works and why I used the word "magic" here.[11]

And what do you think of "magic"?

I would claim that magic teaches one to deal with the unknowable without asking after the "why." A magical consideration would be: How do I learn to deal with things when I have no idea how they function, when I have no idea how I can explain them, and so on? Furthermore, I claim that the art of a magician lies in organizing series of events that are inexplicable. He just knows how to deal with such happenings. Claim number three: We ourselves perform magic constantly. When we begin to speak and assume that the other understands what I'm saying, then this, according to my third claim in "The Magic of Language," cannot be explained in ordinary ways. Take our three-way conversation: We are moving, sometimes falteringly, sometimes disjointedly, sometimes by surprising turns, sometimes in place—*magically*—toward an end. If I am a good magician, I can use this fundamental analytical inexplicability, I can play with this problem. I have it fully under control, without knowing how it works, and thus I am saying that the magic of language lies in its inexplicability. Of course my listeners instantly assume that I'm going to solve the secret of language by devoting myself to magic.

So much for my initial reflections on language. If we now consider the problem of language analytically, at the time our question was this: Can I develop a user-friendly information system in which I can communicate with a machine in natural language, without learning COBOL and without having to master various computer languages? How would I have to fashion the inner structure of this kind of machine, how can I realize semantics in a machine so that I'll be able to get an interplay going between a questioning person and an answering machine?

There was already this amusing dialog-machine developed by Joseph Weizenbaum, who built the interesting ELIZA program, that briefly threatened psychiatrists' raison d'être.[12] But programs like ELIZA didn't interest us very much after we got to know more about them: "Well, a dialog program is very charming, but it doesn't touch on the problem of language, which does not consist in keeping up a game or a conversation." Paul Weston, one of the best people in the area of language according to our way of thinking, worked intensively with the problem of language relations and achieved astonishing results, especially with the phenomenon of semantic structures. He published a work that is very much worth reading in which he really elegantly developed the problems of calculations in semantic structures of relations.[13] The sum of Weston's work: Grammar can submit semantics to a structural restriction, but it can also lead to semantic enrichment. Grammar is a filter through which a rich semantic structure must pass in order to be able to manifest itself.

You made an important analytical contribution in the seventies, which is stressed in "The Magic of Language," which was that the problem of language became fully integrated into the magical game of recursions, and even object names were taken into the realm of behavioral competencies and eigenvalues.

Naturally, it begins with a stable behavior in some given area, as we have already mentioned very briefly. If I encounter a restriction of my motor functions, something that blocks or stands against or objects to my actions—I try to stretch out my arm and can't do it because something stands against it—then I call this objecting thing an object. I can work with or against this object and say, "Well, now how can I go further, how can I win back my freedom of movement? What should I do with this object that resists me so that I can develop a 'better feel' for it?" One begins to manipulate the objecting thing, to manipulate, to manipulate . . .

The hands are back in the game again—mani-pulate . . .

And eventually I know exactly how I ought to operate with this object. I take the stone, lift it and role it up the hill. Then I put on my glasses and observe it rolling back down. And then I can begin the manipulation again from the beginning. It is through circular activities or recursive operations that a stability is achieved with regard to this obstacle that blocks me—and I can also call this "competency." If I know exactly what I have to do in order to "stone," then I call this object with which I "stone" a "stone"; as soon as I know how "to glasses," then I dub this object "glasses." And this is the origin of the remark that the names of objects are actually names and symbols for movement competencies—objects as signs, symbols, "tokens" for eigenbehaviors.[14]

This reversal directs the view away from objects and toward behaviors. And people like Francisco Varela and others who concern themselves with similar language problems considered this turn an original contribution.[15] Not everyone sees an object as just a symbol for a motor competency; instead, we are in the habit of looking at things the other way round. But if you take this point of view then it becomes clear that it is you yourself who is occupied with the stone—and that the others whose freedom of movement is hindered by this object are also hissing and grunting "stohn."

If one has various symbols, the problem of arranging these symbols quickly arises, and it falls to something like a grammar or a syntax to join these symbols together.

Well now, here I'd like to keep several areas apart. I don't think that the building block metaphor that is hiding in the background of your question—"I'm busily gathering building blocks, oh no, now I have to put them in order or I'll lose track!"—describes our issue very well. It is not the case that one starts by gathering up a lot of "stones," "glasses," or "books" as names for motor competencies—they all grow together like a tree. And grammar works like a filter through which these relations are clarified. If you put a comma, then you know a new part of the sentence is coming; if you say "the," "a," or "an," then you tend to expect a noun. One might say a grammar issues announcements for elements that are to follow and the announcements for their part "force" suitable components to follow. In this game, grammar takes on the role of a structural filter through which I pump semantics. But that is only a figure of speech—at this point linguists use completely different descriptions. I'm talking now about the ideas that we developed at BCL to deal with a certain category of problems so they became constructively analyzable. Some of the linguistic analyses, especially Chomskyan grammar and its generative chains, we rejected.[16] They didn't work for our problem. Our critical position regarding conventional linguistics, which didn't seem useful to us, won us a great friendship with Ernst von Glaserfeld. When it came to language problems, he was the very man I always turned to.[17] If someone asked me, "Dear Heinz, tell me something about language," I would say, "Sure, I'll call Ernst von Glaserfeld right away; he will tell me everything about language."

I can only present the collected trinkets that, for me, represent the language problem and the form in which I would like to speak about it. That language is a collection of a society's eigenbehaviors seems to me to be a plausible starting point. I like to think along the lines that our vocabulary, the way we speak, etc. are the eigenbehaviors of recursive systems, that through constant repetition, repetition, repetition . . . this grinding down of language is produced. In this way I can easily understand that just a few kilometers further north or south the eigenbehaviors are slightly shifted because they are always being polished by their users. The users polish and polish and polish—and so local and social differences of usage come about.

You used the beautiful metaphor of language as tree: "And green the golden tree of speech." If we look at various living trees, we can easily see that there are completely different types. There are on the one hand languages in which the word endings are fixed, as in English for the most part—and then there are languages like Latin that inflect; there are language-trees with a fixed word order, tree-groups with a free one . . .

There's a really lovely term from Gordon Pask, "nucleation." Self-organization can develop very quickly if there is a nucleus around which an organization can develop. In my essay on self-organization, to a small extent, one can also find this idea—or rather, invent it within it: We have little wires with hooks on both sides, our hooked-universe from the second conversation.[18] The hooks can take hold in different ways. If, however, I shake differently shaped hooks, then depending on their form completely different structures emerge. The form of the elementary building blocks influences the possible forms of self-organization. And by analogy this also accounts for potential organizational nuclei in the process of nucleus formation itself. Which of the elements becomes the nucleus often depends on chance, on an accidental constellation. I see the different developments of language analogously. Different nuclei create, in part, very different stabilities, different eigenbehaviors. To me this form of language dynamics and language differentiation seems very convincing.

For decades now, there has been this claim buzzing around that there is a uniform plan that is common to all languages. You have already briefly mentioned your skepticism toward Chomsky. Could you explain these reservations in more detail?

The idea of a language organ that produces language is, for me, a typical Batesonian explanatory principle. Just as with instinct, the language organ is an explanatory principle par excellence. The introduction of a language organ manages to eliminate the problem, since the organ produces, well, language. My questions go several steps further: If there is a language organ, are there maybe also organelles that produce numbers? And under the circumstances, would we not also find an even smaller supplementary suborganelle that adds and next to it another that multiplies, and so on? With the language organ we seem to solve only the problem of language: We speak, so there is a little organ that creates language and speaks. That's lovely, but I am not really happy with this explanatory principle—or with any other such principles.

But what's nice is that, well, we can get rid of explanatory principles—Listen up! Here we come to Foerster's razor: "Explanatory principles can be eliminated without loss." I can also throw the word "instinct" out the window because it is so successful and explains every behavior; I can escort the language organ, along with its organelles, off the premises because it explains the creation of language in such grand style. It's no critique from me; it's just to do with our game box and its possibilities—and that is not my game. I play scat, I play bridge, but I do not play with organelles or language organs.

For today's conversation I've brought along with me a book that fits in here because of its title: The Language Instinct.[19]

Ah, that's fabulous, how beautiful, that's exactly what we need.

*At one point the book reads: "Spiders spin spider webs because they have spider brains, which give them the urge to spin and the competence to succeed. Although there are differences between webs and words, I will encourage you to see language in this way, for it helps to make sense of the phenomena we will explore."[20] Through such sentences Molière's scholar smiles at us, the one who 'explained' the effect of a sleeping pill with references to its effect—*quia est in eo*—because the effect is within it.*

Thank you very much, now we can chat about something else. It's so beautiful, and I find it so funny, that's just splendid. "Somebody finally got it right"—it's even on the cover!

Let's continue our conversation about language using language—and let's move slowly, on the lookout for the bumps that we generally run into when we hit its limits.

All right, then, on with language about language against the language bumps! First I'd like to bring in a footnote and show you this little list. Language takes on another structure that depends on how I myself behave toward the world. Am I an observer who stands outside and looks in as God-Heinz, or am I a part of the world, a fellow player, a fellow being?

Language
Appearance and Function in Contradiction

Appearance	Function
World and I: separated	World and I: one
Schizoid	Homonoid, monoid
Monological	Dialogical
Denotative	Connotative
Describing	Creating
Tell it how it is	It is how you tell it
Syntax	Semantics
Self Obsession	Participation
Solipsism	Omnipsism
Cogito, ergo sum	*Cogito, ergo sumus*
Consciousness	Conscience

This list touches on those two areas that the linguists call connotative and denotative.[21] The idea of the denotative is that you refer to something and say "bench"; you point at something and shout "table," and so on. In this perspective, language works like Pavlov's dog: To start with, someone points at something with their finger, you hear a hissing, grunting sound, and you understand, you salivate mentally—until after however many repetitions the finger is no longer necessary, the hissing, grunting sound comes—and you've got it. On the other side there's this idea: You create sounds and hope that they trigger suitable semantic relations in your conversational partner that will engage in relations with your semantics. These two worlds stand opposed to each other: Connotative versus denotative.

I don't always see such sharp oppositions between the individual pairs. I can find many places in your work where, as a participant and co-player you describe—and don't create. Today we've already talked about how semantics and grammar form close, very close interrelations, and the concept of grammar that you and I have used includes syntax. Here syntax and semantics stand in the sharp opposition that we have actually already resolved.

The thing is that one constantly jumps back and forth and oscillates between these areas. If you are talking about language now, you gallop along one line, then you gallop along the other line, and that's the treacherous thing about language, just when you've started to feel at home on one side, language is already standing on the other side again. Imagine the story of the tortoise and the hare, just the other way around: you run and run—and the tortoise is always already gone. . . . For me the most important distinction in the table is between "Say how it is" versus "It is how you say it." These for me are the really fundamental differences between "standing outside" and "standing inside"—and here, of course, syntax fits as a set of rules that you can see from the outside. Semantics, however, is like a roast that is being prepared and will soon be served.

In the genre of a western, the distribution of roles would be clear: The good guys are in the community on the right side—dialogic, ergo sumus. The bad guys form a monological band of solipsists in the left column.

I don't want to distinguish between good and bad with the two columns. Well, one could perhaps interpret these subtitles in that direction. On one side, "connected with the world," and on the other side, "outside of the world." And if one considers oneself to be outside of the worldly or to have internal world-connections, then perhaps the other side can seem to be bad, but it isn't like that at all.

How would you comment on the distinction between consciousness and conscience? I, for example, would like to be consciously conscientious.

Let's assume that we want to make these distinctions and to differentiate between consciousness and conscience. In the state of consciousness I would just say: I'm sitting here and talking, my friends are sitting there smoking cigarettes. I am conscious of these events of myself-sitting-here. Now comes the question: Wherein lies the difference between conscience and consciousness? At this point I appeal to my conscience. A question will be put to me that I must put to myself: "Who does this Heinz actually want to be when he answers?" So I see the conscience as a second order of problems, so to say, in which I talk about how I would like to be. Consciousness, according to my suggestion, does not talk about how I would like to be but about how I am: "I am looking at a bird right now," "Now we're talking about language," and so forth. For me this form indicates that conceptually consciousness moves on the first order. I can, of course, ask, "What is consciousness?" "Am I conscious of my consciousness?" But then, in my opinion, questions of conscience come into play. Here I begin to ask myself, "Who am I actually?" "Who should I be—who do I want to be?" I hope I've managed to show why I move these two concepts along different tracks and divided them into different columns.

In our first conversation, however, you said that the expression "consciousness" is, to use a coinage from Friedrich von Hayek, a typical "weasel word." Everyone uses a concept, but everyone in a different way. I can just imagine a book by twenty-four theologians, psychotherapists, and social psychologists on the theme of "conscience." What sense does it make to operate on both sides of the world approach with these kinds of wavering conceptual shapes?

In normal usage, these concepts are totally familiar. Everyone gets the picture if someone says he was unconscious or, "I see you're very conscientious." In the everyday these expressions work wonderfully. The situation is very similar to the famous Augustinian paradox: "If no one ask of me, I know; if I wish to explain to him who asks, I know not." It's exactly the same with "consciousness" or "conscience." In everyday speech we talk without difficulties about consciousness in *this* context or about conscience in a *particular* situation. If someone asks me, "Tell me Heinz, what is consciousness?" then I've got to answer, "Questions about 'consciousness' give me an opportunity to find out about the person telling me what 'consciousness' is." Afterward, I won't know any more about what consciousness is, but I will know more about the person who thinks he knows what consciousness is. I claim, I believe, I feel that consciousness belongs to the inexplicable or the nondefinable areas. Of course, definitions in a dictionary style, those I can write, but if we want to know how it works, why this system is conscious, that I can't possibly say, it's an undecidable question that is also directly connected to nontriviality.

The Latin *con-scire* is important; it means, "knowing together." Now there are two interpretations of this knowing together: The one is that my entire sensorium "knows together," that means hearing, taste, sight, and so forth, all these sensory streams are "known together," it is *con-scientia*. And the other interpretation is that we are all sitting here together and—*con-scire*—know of each other. In this sense a distinction can be drawn: The "knowing together" of an individual is consciousness, and in a group it's conscience. In both cases, a togetherness is suggested, in one case as a running together in the individual, in the other as people hanging together. Conscience: I am conscious of the other. Consciousness: I get a unified view of my senses that makes me conscious of what is the case here and now.

We'd like to use language to talk about language traps. Some warning signs can be put up here already on the level of words. Some concepts seem to be directly subject to the requirement that they be given warning labels in the form of "Warning: The use of these words may be damaging to your mental health." We've just touched upon one such word, "consciousness." But where "consciousness" is, "I" is never far behind . . .

"I" is very dangerous . . .

Could you perhaps say something more about the danger of this singular personal pronoun, particularly since it was deemed irretrievably lost more than a hundred years ago?[22]

Well, now, whether a word is dangerous or not always depends on which standpoint one takes. For people who fear the development of paradoxes in any form, "I" is an extremely dangerous matter because "I" contains a fundamental self-referentiality. Historically, the paradoxes of logic arose out of self-referentiality—as variations of "I." Take someone who claims of themselves, "I am a liar"—that, essentially, is how the famous Epimenides paradox goes; the Cretan Epimenides says, "All Cretans are liars."[23] And now all at once people were at a complete loss: If he's lying, then he's telling the truth because the Cretans are all liars. If he's telling the truth, then it's a lie and the Cretans are not all liars. What is going on here? This paradox, I would claim, led Aristotle to accept only propositions that could be identified as either true or false, so that a sayable proposition must be either true or false, a principle that has since then established itself in logic and has been maintained.

As far as I can remember, in the Middle Ages the various logicians delighted themselves by inventing paradoxical situations. For me the most absolutely beautiful paradox is about the poor barber—no doubt you know it. A barber in a small town or a small village like Pescadero shaves all the people who don't shave themselves. So far, so good. Why should he shave people who shave themselves? But now self-referentiality comes into play when the question arises of how the barber deals with himself. If he shaves himself, then he's not allowed to shave himself because he only shaves those who don't shave themselves. If he doesn't shave himself, then he has to shave himself, and so on. I ask, "What's the poor barber to do?" These paradoxes were examined more and more over the course of time, but always with the desire to get rid of them once and for all.

What did poor old Bertrand Russell do?[24] Russell struggled with paradoxes like Siegfried with the dragon: A tiny point, an inescapable remnant of vulnerability and susceptibility remained. Russell said to himself, "For God's sake, if I'm writing *Principia Mathematica*, then all the paradoxes must be solved once and for all." Now, he couldn't eliminate the paradoxes; he sat there, thought more, sat there, thought even more, kept sitting there. . . . In his autobiography he describes this time beautifully—every day after breakfast he resolved to solve the problem till afternoon. Paradoxically this situation kept repeating itself—until Russell escorted the paradoxes off the premises: "This is going too far. One of the most intelligent people in one of the most intelligent nations of men cannot spend so long dealing with one of the silliest logical problems, I will simply forbid paradoxes." Well, he did forbid them when he, Bertrand Russell, decreed that one may not use any propositions that refer to themselves—no more Cretan who can flat-out call all Cretans liars. The new form, when Cretans talked about lying Cretans, consisted of the construction of a metalevel and a metaproposition that enable me to talk about the proposition on the lower level. And with this the famous hierarchy of types was born, and ever since Russell we must climb up and down between the first type, second type, third type, and so forth.

My dear friend Gregory Bateson did unfortunately fall down—or up—Russell's steps and was always talking about theories of types and systems of types, a step higher, just a step lower, up and down. At the end of my history of paradoxes we return to the "I," and in my case Heinz von Foerster who claims, "It's all—to speak plainly—an excuse and a hoax, a 'Schmäh' we'd say in good Viennese." All these people just didn't want to deal with people being allowed to speak about themselves. For me, "I" is a folded up recursive operator of infinite depths, but one can operate with it, one can operate with it without any trouble.

After Russell and Gödel paradoxes were analyzed better and better, and in my opinion Ross Ashby already saw the "definite" solution, but it was explicitly formulated by G. Spencer Brown.[25] I can paraphrase the Ashby formulation like this: Every doorbell is, according to Ashby, a paradox because when I push the bell, electricity is sent through a magnet, the magnet attracts the clapper, and the bell sounds. As soon as the magnet attracts the clapper, it has interrupted the contact, and the magnet lets the clapper fall. As soon as it lets the clapper fall, the magnet works again, attracts the clapper, it goes "bing"—and in succession, "bing bing bing bing . . ." The paradox simply solves itself in a dynamic, in a stable dynamic. In his very lovely and important book *Laws of Form*, George Spencer Brown consistently developed the idea that paradoxes create time, a very amusing and clever formulation that I really like.[26] But Ross Ashby had already grasped exactly this point because in paradoxes a dynamic develops in which one condition creates the other and the other the one. From a dynamic view, the paradoxes of self-referentiality don't represent any great problem, and the fear of falling into self-referentiality needn't be met with an infinite series of steps and stairs. By letting ourselves in for paradoxes we find ourselves in a dynamic game in which the one creates the other—and vice versa.

Another language trap is represented by nominalizations such as "thought" or, from the perspective of philosophical history, the amazing gravitational pull of the nominalized infinitive "being."

Very much so, wonderful. In American grammar we do call this process "nominalization," although as a rule those people who use nominalization are not nominalists. The magic of language: "Voilà, a noun is pulled out of the nominalization top hat." In this way a process becomes an object, and as soon as one has completed this transformation, the question immediately arises, "Where is this new object, what does it look like, where does it sit?" Horrifying traps open up here, for example in psychology where one looks for "memory" and believes that this function must have its own chest, a special filing cabinet because I happen to be able to use it as a noun. Indo-European languages especially tend toward, virtually invite, the creation of such traps. I don't believe that these traps exist at all in other language groups. I believe that in Taoism, in the Chinese philosophy, no nominalization is possible, and therefore these kinds of problems don't arise there. I'd like to study this point more closely, for the moment I'm just making a claim.

Perhaps the most subtle language traps are those blind spots where language goes on holiday for no reason. Some of the most richly informative language traps are revealed in the stories of split-brain patients who have had the connections between the two hemispheres of the brain cut for medical reasons. In one test setup by Michael S. Gazzaniga, such a patient sees two pictures—a winter landscape with the left eye, a chicken claw with the right eye.[27] The hands select two suitable symbols for the two pictures—so a shovel for the winter landscape and a chicken head for the chicken claw. Now, the winter landscape perceived with the left eye is processed in the speechless right hemisphere, whereas in the left hemisphere language and visual processes cascade recursively. What happens? A story is immediately invented that connects chicken head, chicken claw, and shovel. They say, "Well, I've seen the chicken claw, and the chicken head fits with that. And therefore I have chosen the shovel because . . ."

Chicken poop . . .

Right. "The chicken poop has to be cleared away—and for that I need a shovel." Here language proves to be a fundamentally unknowable trap because it functions as a great interpreter of what happens in our enormous neural network—sensory and motor.

This reminds me very much of the circular questioning used in family therapy. In this method, questions are invented that initially were not asked at all by the so-called patients or by the people who are now seeking help to solve their family problems. They suddenly face these questions and at first they don't know how they should act: "I'm paying for my therapy, so I've got to have a quick answer to hand." But the question is challenging them: "What do you think that your daughter thinks about her relationship to your husband?" They've never seen the problem from this side, and now they invent, produce an answer. The immediate invention of structures of relations in just these unusual cases seems very important to me. This invention of bridges happens all the time in everyday life, for example, when one has to tell a story. Therein lies a fantastic possibility for new connections, but also for dangerous traps—for example, if one takes these relations as absolute.

With that we're led back to an important point, namely Bateson's "the pattern which connects."

This very important principle in Batesonian philosophy, the search for the pattern that connects or the command, "always mind potentially connecting patterns," runs all the way through the book *Mind and Nature: A Necessary Unity*. Right at the beginning Bateson writes that he's offering a paraphrasing of his ideas: "the pattern which connects."[28] I once thought about a possible extension of the Batesonian connecting pattern. "Pattern" comes from "papa," "*pater*"; it is the father whose stamp is printed across everything, always looks exactly the same—that is the pattern. This I would like to compare with a mother, a matrix—"the matrix that embeds." In this way I've gained parents, the papa, the pattern that connects, and the mama, the matrix that embeds. The papa is responsible for the pattern, and the mama makes sure that the pattern falls on fruitful ground in which it can blossom, thrive, and spread itself. The pattern needs a breeding ground, one might say, that is maintained through the womb, the matrix, the mater, the mama. I feel these two poles belong together, the Batesonian pattern and the Foersterian matrix—I'm always thinking about women, aren't I?

Some slight protection against language traps of all sorts could be, as a little detour in our conversation about language, that one encourages the field of learning to learn. Bateson also made an interesting remark on that deutero-learning.[29]

First off, it's important to see that one can learn to learn, that's important in and of itself. As soon as such a concept arises, seeing a learning possibility in learning itself, then one sees the whole problem of learning in a completely different perspective. For example, I can then use a certain topic, let's say, the history of Charles the Great, as a tool and not as an object to be learned. I no longer ask, "When was the coronation of Charles the Great?" You no longer orient yourself toward the topic, but rather take the topic as an object with which to demonstrate learning. This step, all by itself, is so important in my opinion because using this second-order perspective we can suddenly recognize problems that we didn't perceive before.

With learning to learn my participation takes on another quality than when I have Charles the Great's coronation date drummed into me. I suddenly get a feel for why I'm doing all of this, why I want to know it, why something is happening, and so on. I form completely different connections to that which I'm learning because the object is only a vehicle, a tool to direct my attention to learning. I think that this quality, learning to learn, being able to learn to learn, is something that children have by themselves—they learn so swiftly, they know how one learns to learn. They don't just know how to learn to speak but how one learns learning to speak. That's why children's language acquisition happens so fast, the children have no idea how they do it—it's precisely because they learn learning and not speaking.

As before, we are still talking about language with language—and in a final turn we'd like to approach the problem of scientific language in an even more self-referential way. One of the great ironies is that over the years and decades the two leading theorists of language and learning, Noam Chomsky and Jean Piaget, have not wanted—or been able to—create patterns or embedding contexts that would have connected the two aspects. As far as the pattern and the matrix go, they remain orphans.

Here I've got to turn at once to Massimo Piattelli-Palmarini—a brilliant, young, funny, lively, and clever lad who managed to get a fantastic castle, the Royaumont abbey north of Paris, as the location for a conference and invited the two giants of the theme "language and learning," Jean Piaget and Noam Chomsky. And grouped around the two of them were various people, the Harvard and Yale IQ-superkids, whose arrogance is so overwhelming that one never ceases to be amazed, Nobel Prize winners in biology, philosophy, psychology, anthropology, cognitive scientists, they were all stationed in the abbey. The theme was "On Language and Learning," and that's also the title of the anthology that came out of this conference, which for me is one of the most important books of the twentieth century.[30]

What's fascinating is that the conference participants managed to sit together for days and *not* listen to each other, not listen for even a moment. To the question "In general, what do you think about mutual relations?" would come the answer, "My phone number is 5679 plus the area code," to which the first would respond, "That's fascinating, but now let's talk about the complex proposition 'All swans are white apart from those that Popper labeled as black,'" and so on and so on. It was fantastic, endless entertainment. Piatelli-Palmerini describes this form of monologism very well in his foreword; one simply has to read between the lines. The great thing for me was that Piaget was able to present the central ideas of his work in a few sentences, although nobody listened to him, nobody went into it and no connections were made. For me, Piaget's explanations in this volume are very, very significant. There is a wonderfully dreadful movie, *King Kong vs. Godzilla*—and this conference bore a great similarity to it. King Kong was Chomsky and Godzilla Piaget. Bärbel Inhelder, who was still alive at the time, would try from time to time to say, "Would you listen to this person for a moment," to no avail.

This argument about language and learning does, however, tell us an unbeliev-able amount about the functioning of scientific languages and, above all, about their limits.

Right, these great men managed to apply neither their very own specialty—language—nor their core competency—the learning of theories—to their dialog. Instead, they dispensed with their facility for language because they were following some pecking order, defending some preserves, were interested in connections with expensive universities or publications or simply wanted to maintain their status as arrogant super-masters of the mental air space. All of this and much more stood in the foreground, rather than them going into the problem of language for even a moment.

Even Bateson himself, who was present at this conference, wouldn't find a move that would bring a connecting pattern into the debate between Chomsky and Piaget.

Bateson, who was such a lovely man personally, was so overwhelmed by this wave of arrogance that he could hardly open his mouth. Piaget, on the other hand, pulled through the confrontation, unwavering as a Swiss peasant, and didn't let anyone confuse him, not Chomsky, not Fodor, not Putnam, Thom, or whatever they were all called. It's just that no one listened to him. Some went outside and smoked cigarettes—a grand circus, a dreamlike place.

The Piaget-Chomsky conference is an example of a recursive production in which, however, no eigenbehaviors, no consensual domains peel themselves out. You already mentioned some of the reasons why that happened—the arrogance of some participants, the claims of some top dogs. What are the possibilities that a recursive event won't move from the spot?

Intentions, for example, play an important role: "I just don't want to understand it," "I don't want to engage with this topic," and so forth. There are also affective components that exclude stabilities for convergences in scientific discussions: "Just looking at this person makes me see red," and so on. I would just like to point out that these phenomena are not limited to the sciences. Such people can't shovel a path together, can't build a pump together, can't make a fire together. It just doesn't work. They are, if you'll allow the image, two different chemicals that don't react with each other.

In formal theory there exists an interesting theorem on consensus-building processes in groups, which was formulated by Lehrer and Wagner.[31] If a group has reached a certain stage in which everyone is committed to a certain opinion, consensus may nevertheless be reached through an iterative process, as long as no one in the group sees the opinion of another as totally irrelevant and weighs it at zero—incidentally an interesting variation on your "one can learn even from the dumbest." This result represents in a certain way a demonstration of the magic of recursions—and the power of dialogs.

My interpretation of the theorem is as follows: If, to put it in my language, one element of a process of consensus building or even in our dialog or trialog doesn't want to play along, no convergences will arise. Now you will also understand why I don't like this paradigm idea and why I'm sometimes unhappy if my dearest friends use this vocabulary. That, for example, the dear Fritjof Capra constantly rides around on paradigm shift like it's a bicycle does irritate me a little sometimes.

The inventor of the paradigm shift, Thomas Kuhn, did, however, give a very intelligent description. A community of scientists endorses a very particular idea, allows itself to be convinced, is convinced—and spreads this new perspective.[32] What's important here is that the persons involved must make an active contribution. The concept of paradigm change—used in the sense of its creator—is also something we could actually accept.

I like Thomas Kuhn. His idea of paradigms brought to light the blindness of scientific theories that could not or would not understand such processes. I find this part of his explanation splendid. From one of his ideas, however, I would like to keep my distance: Kuhn says that if a paradigm ceases to function in a variety of cases, then a paradigm shift is on its way. I claim just the opposite: Again and again we can show that once a paradigm is brought to perfection it topples and vanishes from the scene.

Let's take, for example, the Copernican revolution. Tycho Brahe could work out the solar and lunar eclipses wonderfully with his epicycles—the ellipticians, on the other hand, couldn't do that. They only had a fixed parameter for the definition of an ellipse—and could not in this manner calculate the orbit of Mercury at all because Mercury doesn't follow an elliptical orbit. The perfection of the earlier paradigm, I'd claim, and not its faults, its failures, mobilized the countermovement—"That's getting too boring, always the same epicycles, couldn't we interpret it differently?" I know it sounds heretical: The perfection of the old paradigms, not their flaws, contributed to the establishment of a new view, which, although it was not at first as mature as the old one, was however accepted with enthusiasm by many people, especially the young who played along with it. Aesthetic considerations also play a great role here: It was more elegant and less boring to calculate with ellipses than with epicycles. And then finally Newton comes onto the scene and says, "I don't make hypotheses, it has to be this way." Actually, it was a great succession.

Let's come back again to the concept of the paradigm, which is less than valued by you. What, then, is the reserve concept that you would rather use?

I like the expression "style" or "thought style" better. The styles of science are not fundamentally different than the styles of draftsmen, of painters or of literature. Styles don't necessarily use force however—thank God the protagonists are sufficiently diverse, in living organisms there are no exact copies, no precise repetitions of sameness.

This concept of "thought style" emerges for the first time in the work of a Polish physician, Ludwik Fleck, in the 1930s.[33] Fleck always uses this concept of "thought style" in conjunction with a collective. An individual cannot have a thought style; only a community can.

Absolutely right. A style of thought belongs to a community—and it develops through the formation of commonalities. Such commonalities aren't there from the beginning; they arise through playing together, in ping pong games, in language or thought games—as we go along.[34] Only through this does a style of togetherness emerge. All participants are different, each is and remains another; only, in exchange and interaction does a "style" arise. It emerges. We could speak of the emergence of a style.

A single, innovative idea is then never the work of an individual.

That's absolutely clear. If I come back again to the example of my life: Very diverse and controversial persons worked at BCL, but on certain questions they would mutually stimulate and spur each other on. Even today our work together at that time seems astonishing: Such abysses gaped between Ross Ashby, Lars Löfgren, Humberto Maturana, and Gotthard Günther; and yet they talked, discussed, and quarreled with each other—and could come to an agreement time and again.

A few of these outlines correspond to certain styles of thought that are very quickly disseminated, and a special class of scientific products have a particularly hard fate—they come too early, are too radical at first, too incomprehensible.

Science and its products have to fit into their surroundings, to return to a Glaserfeldian expression. And the surroundings, this environment, are in a certain way also politically oriented. Why? If you look at the history of new ideas, then you constantly find innovative ideas and proposals that couldn't gain acceptance. The political climate, to use a meteorological image, was such that these ideas were suppressed by other ideas. We've experienced this "hard fate," as you call it, only too often in our own lives. Our idea of "parallel computation" was totally incomprehensible at the time. Colleagues said, it's pointless, that we just didn't know what we actually wanted, and so on. No matter how enthusiastically we talked about it: "Just listen, in a parallel machine, like the eye for example, there are about 10 million operators, namely the rods and the cones. They operate very slowly; each of these lazybones takes a whole tenth of a second to react. In spite of this I can recognize my uncle faster than any supercomputer today—because we do everything at the same time and compute everything at the same time. If 10 million look for a tenth of a second, that makes 100 million operations per second."

People didn't understand that. We've seen it again and again, that someone will come along with some idea that it only becomes possible to realize later on. Today "parallel computing" and "connectionism" have become perfectly natural. It was fascinating to see a milieu in which one could set out a thought, and no one would know what to do with it, it just couldn't get a toehold, it—I'm changing to a botanical image— couldn't put down any roots. I tried at the time to summarize our ideas in a detailed article, "Computation in Neural Nets."[35] In it I showed what complications could arise, ways of calculating in a networked way, which methods there are and could be, which operators are necessary and how one can interconnect them. That was all found in seventy or eighty closely printed pages that appeared as the main article in a new journal. What you find there in principle became a main subject twenty years later: "Parallel Distributed Processing."[36]

What's remarkable about this article is that no one cites it although the basic ideas are constantly being taken on and borrowed. McCulloch once said to me, "When one is young, then one is dismayed or enraged if one's ideas get stolen. When one is old, one is pleased and feels honored."

Because we're getting toward the end of our conversation about language, in which we are now dwelling on the area of scientific language in terms of scientific language, let's consider a further problem: namely, the widespread fashion trends and booms that also manifest themselves in scientific language.

I would call that PR, public relations, foreign relations if you like. You've got to manage to invent a word that suddenly becomes widespread in a social milieu. Take, for example, "chaos" instead of the term "recursive function." In the case of chaos, a rapid diffusion took place; with recursive functions, hardly anyone listens. Or throw in the term "catastrophe theory," and immediately everybody finds it terribly important, and all the newspapers write that René Thom invented catastrophe theory![37] For a long time this field was known as "bistable systems." The first equation on bistability comes from the nineteenth century and has to do with condensed gases. What later became so suddenly famous as catastrophe theory could already be observed and formalized in the nineteenth century: "Is a gas liquid, or is it already steaming?" It cannot exist in both states; under certain circumstances it goes from one condition to the other. These thermodynamic states were already recognized, understood, mathematically expressed, calculated, applied, and graphically processed—there were for many decades these beautiful discontinuous S-curves, on which one could just jump from the bottom to the top part of the S.

If you find the right words, people take notice, the sponsors, the agents, the financial backers. Everyone likes to support chaos theory because they want to better understand a blind, broken world. Everyone eagerly promotes catastrophe theory. Everyone feels drawn to attractors. But bistability, recursive functions, eigenvalues, such expressions apparently sound too traditional or theoretical. And it didn't matter that the simple questions of the Apfelmännchen had nothing to do with either chaos or disorder—nobody cared about it.

Let's go through some of the key concepts that have come up in the course of our conversations over the last days a little more systematically. And since we're already at "coming up," let's start with "emergence."

I allege, a little maliciously, that the emergence of emergence owes itself to the desperate search for new research money, for sponsors and patrons.

The problem that emerges for me in the case of emergence is a typical one from farming: Can I milk the cow better with this concept than my colleague who it so happens has not yet fallen for the beautiful idea of emergence?

Let's take ourselves back in time some decades, to the age of bionics research, of bionics.

Bionics was once one of these terribly fashionable ideas. Rather by chance, I myself was involved in the bionics field very early on because I knew a colonel in the Air Force who was hatching this funny idea of bionics. At the time the Air Force had difficulties getting research money from the American Congress. And then my Colonel Steele managed to send his magic word on a journey—"bionics." And immediately Congress turned around 180 degrees: "Aaah, 'bionics,' unbelievable! Who is the best bionics-man we can find, where is the best bionics laboratory?" It so happened there were two or three people—Warren McCulloch, Heinz von Foerster, and a couple of others—who were working in the field of bionics at the time. Naturally, we didn't know that. Well, be that as it may, we ourselves were very early in the bionics business. For years bionics was a real draw; you really got millions of dollars from Congress for it. I have to say that I was lucky enough to sit directly under the cow's udder that was giving much of the bionics money.

Bionics, however, soon found itself in a quarrel with another militarily attractive vision, namely that of "artificial intelligence."

Right, these people suddenly fell for the ingenious idea of artificial intelligence. Then suddenly the artificial intelligence cow got an overfull udder—and these people started to milk the cow good and proper. These people were even more politically skilled than our bionics troop because they cooperated with the Marines. For reasons unknown to me the Marines always get more research funding than the Air Force—so the Navy started to push artificial intelligence and support it generously.

A key word, one that is less martially and militarily monopolized, is the concept of "autopoiesis," which, despite its whiff of traffic and poetry, became particularly popular in biology and the social sciences.

Yes, in my opinion the concept of autopoiesis was a good invention of my Chilean friends. It's just that—it couldn't get a foothold anywhere, and one got no research money if one was interested in autopoietic systems.

But in Europe, particularly in the German-speaking areas, autopoiesis has become especially important, especially in the social sciences.

In Europe, strangely, it works as a key word. Niklas Luhmann took it up, others used this concept. Nevertheless, to my knowledge, the research-funding horns of plenty don't pour onto the autopoieticists in Germany either. Money is mobilized by artificial intelligence; money can be expected in the field of parallel computer architectures—but you're not allowed to call it parallel computation. The buzzword of the moment is "connectionism." Today, if you've got connectionism on your lips and you construct parallel computers, the entrance gates will open wide for you. When I wanted build parallel computers in my day, I only ever saw the outside of the entrance gates.

One could now diagnose a widespread irrationality in scientific language: If someone had gone around talking about artificial intelligence in the 1960s, they wouldn't have gotten any money; if someone enthuses about bionics these days, they won't get any money either, even though the aims of both groups are formulated very similarly. At times this predominance of pure labels seems pathological. But to a consistent systematist or observer of wholeness like yourself, phenomena such as faddish ideas, conceptual booms and busts shouldn't seem strange or incomprehensible—wouldn't it be much more mysterious if such processes were not found in scientific language?

Yes, that's completely right. When I talk about it, it sounds like there's a malicious intent behind it—absolutely not! These groups chanced upon a concept: "Bam," they got lucky—the concept spreads rapidly and finds its way into elevated everyday speech. Journalists are especially delighted by new labels. They can copy their old articles with the new concept written in.

For chaos theory, the name "chaos" was significant because everyone seems to understand it, because everyone's had to search for something in their living room or office. In the twentieth century it was the theory of relativity, because many were interested in the relativities in their own lives, although the connection between relativity and the theory of relativity represents a typical misalliance. And in field of self-organization or system dynamics, it's unavoidable that verbal dynamics will come forward with titles like Fractal Factories, Management Through Self-Organization, *and so on. Central concepts diffuse into other milieus and create other understandings there.*

In the general understanding of the theory of relativity, everything is relative, while the theory of relativity itself claims exactly the opposite. But in this example we can see very well the differences between theories, naming, and dissemination.

Under the circumstances, did you at BCL perhaps fail to invent one or two catchwords?

We didn't just fail, we didn't even understand. Looking back I have to wonder at myself: "You, clever Heinz, you don't get that it also depends on the label and not just on what lies behind, below, over or within it." At the time I was so very enthralled with the inner richness, with the inner fascination of our research ideas that I forgot to carry these inner beauties to the outside. The people who worked with us were also too strongly obsessed with the object to waste thought on selling and popularizing.

"Radical constructivism," meanwhile, has become a term that does well on the market and has a positive influence on sales . . .

It may be good for sales, but I'm not particularly fond of the expression. If someone says to me, "Heinz, you're a constructivist!" then I answer, "How so, please? What is that? Could you explain to me what a 'constructivist' is?" No one can give a satisfactory answer to that. And furthermore, what is a "radical constructivist" as opposed to a simple or nonradical or normal or nonrooted constructivist? No one can adequately describe what all this is about—but the expression is constantly used.

SIXTH DAY

Experiences, Heuristics, Plans, Futures

Not knowing . . . how one was born, the navel, an ontogenetic necessity, is an ontogenetic riddle, a mystery, or a joke.
—HEINZ VON FOERSTER, *Understanding Understanding*

Each sentence that I write is trying to say the whole thing, that is, the same thing over and over again & it is as though they were views of one object seen from different angles.
—LUDWIG WITTGENSTEIN, *Culture and Value*

And God made the beast of the earth after his kinde, and cattell after their kinde, and euery thing that creepeth vpon the earth, after his kinde. . . . So God created man in his owne Image, in the Image of God created hee him; male and female created hee them. . . . And the euening and the morning were the sixth day.
—GENESIS 1:25–31

As far as I can remember, we're now dedicating ourselves to the mythologies, strategies, technologies, jokes, and so forth, that this Foerster uses to sell his curious intellectual soap bubbles. Is that right?

In our big game box we find countless programs. We are searching for the Foerster modules.

Foerster modules, programs, programmed . . . I'm not happy with using the concept of programs at the moment. Naturally, I do use the concept of programs whenever I hope that an important distinction, a fruitful heuristics will come out of it. People know what the word "program" means. In the case of the differentiation between trivial and nontrivial machines this distinction seems to have functioned well, lots of people have taken on this distinction and have adopted it, lots of people say, "Ah, now I understand more about my environment," and so on.

In our conversations about the magic of recursion, we encountered two types of operators: operators of the first order and operators of the second order, which send the first-order operators out on journeys. Let's try through conversation— perhaps supported by the expression "module"—to find such Foerster operators on the first and second levels.

Before you get into outlining programs, I'd like to warn you of something. There's an important reason why the idea of programs and of Foerster operators is so unappealing to me. Namely, I've noticed, and before now, that I seldom reflect on myself. Though I often think about the "I," I never actually think about myself. I think psychoanalysts would have a hard time knowing what to do with me. If someone asks me, "What do you feel about X?" or "What does Y do to you?" I have no idea! I seem to be rather spontaneous. If the situation is like so, I do this, if the situation is different, I do that. Whatever I do, I don't do it with intent or out of long preparation. It just comes, and then I act as best I can. I don't know if we'll get very far with "programs," "operators," or "modules;" we might not even get off the ground.

Let's sail a little further—at least metaphorically—under the flag of programs. Programs can contain lots of chance components, can prove flexible and reconfigurable! Let's use the heuristics with the first and second level operators—and let's look at operators such as "modules," which will together lead us to the Foerster program, and to get into the mood, let's begin with the first level of operators.

I repeat: I have no plan, no intentions, I don't understand why I do this or that. And that's probably why I stumble across inexplicability in so many areas, because I can't explain or predict myself. If programs and modules help with this, then let's operate with them.

Let's try with the operator that I take to be the most important Foerster module, namely, "inversion," the reversal, the turning upside down of settled, traditional relations. This operator seems a little like a conceptual Jacobin: "Down with the king, long live the revolution!"

Well, that I like better. And that reminds me straightaway of an important episode from the time of my studies. One of my colleagues at the Technical University said to me, "Heinz, I've been to a couple of lectures at the university that you have to go to. There's another one tomorrow, let's go together!" We went to the university together. The lecturer was a professor Scheminzky and the title was, "Can life be artificially created?"[1] I entered the auditorium, and it was already jam-packed. In the front row there sat, naturally, the great professors of biology and the rest of the great and the good. The chairman announced, "Professor Scheminzky will now speak about the problem, 'Can life be artificially created?'" Upon which, the men in the first row all stood up as one and marched out in protest. The group with the respectable beards, the great professors, were just gone. We young people said to each other, of course, that this must be the right way; this lecture series and its contents are the right thing for us. The best propaganda for any idea for me is still: The orthodoxy marches out the door. This lecture series was organized by the Vienna Circle at the time.[2]

Yet another example: You'll probably see again and again that I especially like to turn a process or a relation around if an asymmetry is indicated within it. If I find a conceptual asymmetry in any proposition, then I immediately turn it around and investigate what consequences could be associated with that. A very modern term right now is the "bottom line." The American economy today consists entirely of such bottom lines. The bottom line is found at the end of a calculation, and in it is written, "Those were our profits, those our losses, that's our income, those are the expenses for employees, that's the cost of the chimney sweep—bottom line—negative two thousand, two hundred and seventy six point twenty three."

Everyone stares, spellbound, at the bottom line. And in a case like this I start to invert: "Well, you're always looking at the bottom line with such fascination, very good. Do you know what, I always look at the 'top line.' Maybe something is going fundamentally wrong up at the top. Why don't we all take an intensive look at the top line for once." I enthusiastically perform this kind of inversion again and again. If a proposition goes ABCD, then I'm interested in DCBA. I think that theories of humor claim that such reversals—especially when they happen unexpectedly—form the foundation, the core, the point of jokes and humor. If you like you can say that my central theme is—the joke.

One very instructive experiment in inversion comes from your Viennese days— you turned around the propositions of the Tractatus.

I did have certain difficulties in selling Wittgenstein to my constructivist friends, or at least bringing him a little closer. Why? There are some propositions in *Tractatus* that absolutely cannot be interpreted in a constructivist manner; in a way they're a boxing around the ears for constructivists. Take for example Wittgenstein's "picture theorem," the famous proposition 2.12:
        "A picture is a model of reality."
Ernst von Glaserfeld once said to me, "When I got to this point I put the book down—and didn't read any further. Total nonsense to me—first the world is postulated, and then the picture comes afterwards. We're not taking pictures!"
    Good, now here comes Heinz von Foerster starting his "inversion game":
        "Reality is a model of a picture."
Here the picture becomes the cause and the "world," our "reality," the consequence, not the other way round. And naturally the constructivists are very happy with this inversion because this is how they see this connection as well.
    Does this inversion bring you into contradiction with other Wittgenstein "theorems," that is, do you stumble? No, I'd claim that one can—unless one forgets to invert the relevant sub-propositions—develop a totally consistent philosophical picture if you invert his "world-picture postulate"—and build it up as the "picture-world postulate." I think that the situation here is very similar to that in geometry. I have, for example, always been very enthusiastic about the possibilities of non-Euclidean geometry. People have tried again and again to prove Euclid's so-called parallel axiom through the remaining axioms.[3] The parallel axiom says, roughly: For every plane, in which there is a line *L* and a point *P* that

does not lie on L, there exists exactly one line L' that goes through P and is parallel to L. And for a long time the exciting question for geometers and mathematicians was whether it was possible to use the other axioms to determine points, straight lines and curves such that the "parallel axiom" eventually comes out as a deducted theorem. As has been said, this proof was not achieved; on the contrary, Bolyai and Lobachevski were able to prove that the parallel axiom is indeed independent of the other axioms. But if that's so, then I can also deny it and, together with the other axioms, invent a new geometry that will have to be free of contradictions. Because if it contained contradictions, then these would have long since trodden on the toes of the other axioms, who for their part would have screamed so loud that it would have been noticed at least by an axiomist, right? As so in the nineteenth century the parallel axiom was denied and it was claimed, "To every line there is not just one but any number of parallels"—the Lobachevskian geometry—and on another occasion it was posed that "a straight line has no single parallel"—Riemann geometry. In these ways completely new geometries developed, no one stepped on anyone's toes, no one screamed—new consistent, non-Euclidean geometries blossomed, and people began, for example, to work on the geometry of the sphere or the geometry of multidimensional spaces, and so forth.

I see the situation in the case of Wittgenstein's book of axioms, the *Tractatus*, in a similar way. If one turns around the picture postulate and says, "No pictures, rather, examples" or "Pictures, pictures, nothing but pictures"—what systems build themselves up then? I even think that certain Wittgenstein propositions fit more easily into the inverted version, in which reality becomes a model of a picture.

These reversal operations have become favorite regulars of yours. For example, one of the most important reversals occurred with the role of the paradox. It went from being something that was avoided like the plague and put into neat and tidy hierarchies of type into being something that is treated like an equal and welcomed like an old friend.

Yes, we developed a constructive circle or a creative circle rather than a vicious circle. These kinds of inversions come up again and again, they are perhaps, if you wish, actually a piece of "Heinz methodology." If you have a relation that shows asymmetries—"That is primary, the other is secondary and follows from it"—then I immediately turn the tables and look to see which new pictures emerge. As we've already said, this operation seems to be the foundation of humor, the point of jokes. On the one side stands a fundamental statement, for example the one from Korzybski—"The world is not a map."[4] And then comes Heinz von Foerster: "Bam, the world is a map." Suddenly the fundamental statement is a joke, people laugh—and can build new insights on this basis. I think that it represents essential progress if our "fundamentals" are turned around as jokes and, therefore, become entertaining rather than overwhelming.

And this brings us to the navel. You write that our navel is, for us, "an ontological riddle, a secret, or a joke," in that order.

Thank you very much for that quote. Besides that, I conducted experiments with the navel, which I absolutely must tell you about. I asked children who weren't spoiled yet, that is, ones whose parents hadn't yet explained what a bellybutton is good for. So on the beach, where children were running around naked, I'd ask them, "Tell me, what have you got there, what have you got on your stomach?" "That's my bellybutton." "Well, yes alright, but what is the bellybutton doing in the middle of your stomach?" And I got the greatest answer from a little girl who put her finger on it and answered, "I can say 'I' with it." Isn't that uncanny? That really impressed me extraordinarily: "I can say 'I' with it." An answer to a fundamentally undecidable question produces a creative response.

Let's assume for a minute that your positions and perspectives became generally accepted, how would your Jacobin operator react? Would it start to invert again? Would "Reality is a model of a picture" and "A picture is a reality of a model" be turned into "A picture is a model of reality"? After all, a model has got to be a model of something.

No, no, because I hope that in this situation everyone would be laughing. Because I hope that people would no longer break each other's skulls for the one and only truth: "I have the truth, and therefore you cannot have it because you say something different." I hope that in such a case a relaxation of the relations between people would occur because they could make jokes, because they could just turn around all statements, turn them upside-down or use asymmetries, and so on.

Let's leave our "Jacobin" and go to another important Foerster operator. The connecting pattern goes by the name "Jacob"—that is, Jacob Grimm. An operation of yours that we find again and again is playing with the etymological backgrounds of concepts and expressions.

Yes, definitely, you're right. This game with the origins of words comes from the fact that I myself always feel unhappy using words whose numerous meanings and whose origins I don't know. If I'm using a word and suddenly someone asks me, "Hey, tell me, what do you actually mean with 'electromagnetic field'?" then I don't want to only be able to answer, "'Electromagnetic field', it's what I read in textbook XY on page 4." An answer like that wouldn't be enough for me, and I'd also like to know how one got to "electromagnetic fields," where the expressions "electricity" or "magnet" come from, when and how they emerged, and so forth. If I find that out, then I feel significantly better.

Furthermore, I am often impressed by the insights that reveal themselves if ones goes back to the origins of words. For example I was deeply shocked by the expression "science," not by the German *"Wissenschaft,"* that gave me less of a headache, but by the English "science," from the Latin *scientia*. So I looked it up in the excellent *American Heritage Dictionary of English Language*, the last hundred pages of which contain etymological references. I flipped through to the root word—and there I came upon the archetype, the Indo-European root *ski*—and that means to separate. The essential idea of *ski* is to separate and it crops up in every kind of word possible, like "schizophrenia," "schism," but also in the words *"Scheisse"* or "shit," because you separate yourself from these things, whether you want to or not—and if you look it up in the German *Duden* dictionary, you find exactly the same state of affairs.

When I say in my lectures, "Look, 'science,' 'schizophrenia,' 'schism,' 'shit,' etc., they all belong to the same category of separation materials," I mostly get either gales of laughter or angry outbursts. Naturally I asked myself, "What expressions do we have that run counter to this 'separation,' that mean to unify and to integrate?" In this semantic field I came across the Greek word *syn*, together, or even more significantly, *hen*, one. And with this I have the etymological roots of two opposing schematas of thought: The first, the *ski* form, separates, it aims for taxonomies, dichotomies, operates with "get out, get out, get out"—the other, the *syn* or *hen* form, integrates, brings together, acts with a "come in, come in, come in." *Syn*, that's what people like Gregory Bateson do, with this talk of "the pattern which connects." That's something that everyone should really know by heart. What brings us together, what pattern connects the orchid to the primrose, the frog to the elephant and all four of them to us? These kinds of patterns, I would claim, are also created through magic.

Going against your Grimmean operator, one could of course argue that it has too much entertainment value and too little informational value. It could actually be irrelevant where a concept that I use came from; what's important is just that there's a consensus as to its usages. Etymology can look after itself or be left to the linguists or historians.

Naturally, people can claim any kind of nonsense; no problem, I've got no objections to it. That's someone else's problem then, not mine; it's as if someone said, "Good, I've got a bellybutton, I don't need to worry about why it's there, I've just got it." I see the connection to a word's history in a fundamentally different way. Within a word or concept exist some fifty, hundred or two hundred thousand years of human development, and these are also present in the conversation we're having right at this moment about the bellybutton and the world. Naturally, one can say "So what?" to all that. That, however, is no longer my problem, that's the problem of the person who fails to experience this joy, this pleasure of going back to the pre-pre-prehistory and looking into the etymological depths and abysses. If I see a point that represents the projection of a very long line, and someone else is so fixated on points that they aren't interested in this long line, then there's really nothing I can do and I'm not going to start a fight. If the other says, "Etymology is just a game for linguists," then I'll answer, "If you don't want to join in with the game and have fun, then you can just keep boring yourself in peace!"

For the next operator we have to lengthen and stretch our connecting patterns a bit. But from Grimm we inevitably come to fairy tales—and from fairy tales to fairy tale productions and to theatre plays. Our conversations also consist, strictly speaking, of countless little "minidramas" in which you mostly assume the role of a questioner, opponent or listener—and on the other side you play yourself. With this, for me, we come to a further very important point: You constantly play different roles in "possible worlds"—or rather, in "possible plays." We'll call this point the "role operator" for the time being.

Again, I'm less than happy with the expression, but as for the thing itself, naturally I've got to admit that you're right. Throughout my whole life I've slipped into the most various roles, have played them and won great successes with them, but have also caused great irritation. I enacted an especially beautiful role-play at a meeting on "human invariance"—and thereby temporarily incurred Jacques Monod's hatred. Well, Monod has since died. Once again, the smarty-pants of the world were gathered at this meeting—Monod, Edgar Morin, Jerrold Katz, Jerry Fodor, all these unbelievable IQ high-flyers. And an anthropologist gave a completely enchanting lecture on pygmies and explained how these pygmies solve all quarrels and complications within families through funny little theatrical plays. Pygmy family therapy works like this, so that some of the Pygmies dress up as clowns, as comic, funny figures, go to the family that is having the quarrel, and act out the family difficulties as clowns. In the best case, everyone starts laughing, everyone finds it comical— and the problem solves itself; the family is transformed; another relation has arisen between the family members.

I thought this lecture was really beautiful, particularly as the anthropologist presented it in such a nice way and brought us closer to the game of the Pygmies in a very human way. Another episode from the lecture has also stayed with me: the Pygmies try not to see slain animals as enemies or opponents, and so the elephant is not hunted and killed, no, "the elephant puts itself at their disposal," so as to be eaten by the Pygmies. They don't say, "We've got to kill it!" Rather, this animal puts itself at their disposal as a fabulous elephant steak—all in all a marvelous lecture, which I very much enjoyed.

Hardly had the anthropologist finished his talk when Monod, who doesn't exactly suffer from an arrogance deficiency, stood up and tore the speaker to shreds in a most unpleasant manner. When Monod had finished with his punishing monologue, I recalled an example from the lecture in which the behavior of an "über-father" was represented in a funny way. Then I stood up and said, "You see, ladies and gentlemen, we have just experienced one of those cases that the anthropologist told us about; one of the clowns has taken on the role of the 'übermensch' and has shown us how you bawl someone out"—and so on and so on. Everyone was doubled up with laughter, as you can imagine. Monod saw me later in the aisles and hissed, "You did not understand me." That was my first contribution to the discussion on the theme of role-playing.

This delight in various roles also shows itself in that you, I think, have deliberately sought and created closeness and contact with very different people and points of view. Your fellow players, especially at BCL, were grouped around you in great diversity; which, incidentally, is an important advantage for a research organization.

I think so, too, yes. Let me put it this way: I'm just happy if someone develops a clever idea, builds an interesting model, writes an astonishing program, and the like. I have no interest at all in getting a patent claim or the exclusive rights to some idea—I couldn't care less. If someone comes up with a funny, amusing idea, I'm enthusiastic—and I publish it and spread it: "Have you heard, Fritz here has developed this wonderful idea! Look at what Fritz has hit on!" I can imagine that with this attitude I've got a lot of people to come to new ideas with a sense of fun and pleasure, or just to discuss new ideas: "Come in, Heinz, could we discuss this, I don't understand this point!"

The fun and the pleasure of developing something new totally independent of the problem of who said what for the first time and who drew which conclusions probably very much stimulated my many colleagues. They'd say to themselves, "It makes sense to be active here since my work is acknowledged, valued; it's connected to other works," and so forth. This probably reveals one of my intellectual foundations, not to envy anyone who comes up with something new, rather to be glad that there is something new. "Envy" and "jealousy" are just words to me. So that's my second excursion into the matter of roles and community games.

If we turn from the roles to the plays then our attention is almost inevitably drawn to the great diversity of what you've developed in various types of roles. This diversity of plays and fields—from physics to the social sciences—is something that you've kept up for decades.

If I manage to make a thought-bridge, if I can draw a thread from A to B so that I can say, "Ah, from here I can see what's presumably happening in this area," then a conceptual enrichment, created through language, starts to set in. I've got to use an already existing language to create an uncommon connection in myself or in my listener. And this process of connecting is totally independent of whether we're talking about cells, chromosomes, atomic nuclei, molecules, individual persons, populations or other elements. It's all exactly the same; you have a thought method that tries to create links. That we're placed in the area of physics, demographics, chemistry, cognitive sciences, or physiology has to do with which chance elements we happen to be looking at. Because one happens to be talking about cells, the field is called biology. If you're talking about populations and inhabitants, however, then the area is demographics. What interests me in all this is primarily the connections that one can set up there.[5]

The elements don't matter to me; the only thing that matters to me is the connections; they've got me under their spell. The problems are the same everywhere; it's just the elements that move around. Usually you can't find these kinds of connections on the same levels, if I may say it like that; as a rule you have to go down a floor—"Ah, the same thing is going on here; let's move down again!" Working in this way, it might be that to make the connections I need elements that I first have to invent— and so the different floors and stages come about. Thus, if you ask me why I slip into these different levels, I'd answer that there are no surfaces; everything is equally deep. You just have to know how deep the connections are that explain or describe these phenomena.

That's a witty variation on an important proposition from the manifesto of the Vienna Circle—"In science there are no 'depths'; there is surface everywhere."[6]

Good, yes, splendid! Then we've just invented the Foersterian complementary proposition to the Vienna Circle manifesto.

In science, as in scientific research, the expression "interdisciplinary" has come to the fore and stands for a new, up-to-date form of knowledge production. But this label, namely "Modus II,"[7] represents an approach that seems to have been self-evident for you for years, decades even.

Yes, because I'm simply not disciplinary. A history guy, a scientific researcher, would certainly claim, "Disciplines emerged because people busied themselves with certain problems." So someone or a group of people, it doesn't matter for the moment, sat down and developed a telescope . . .

I would like to claim, historically, that, "Disciplines are tribal societies . . ."

Aha, that's how you see it . . .

. . . with chiefs and enemies, friendly chiefs, enemy tribes . . .

Yes, well, well. I'd like to develop my way of seeing it a little further. My point is, someone is interested in stars, builds themselves a telescope, looks through it and finds all sorts of new objects, makes astonishing observations and the public shouts enthusiastically, "Bravo, fantastic!" or, horrified, "What a charlatan!" And to the question of what this man is doing, you answer, "Astronomy." Other people support or oppose our astronomer, and after a certain length of time an astronomical institute will come out of it. In the beginning, however, there was just this chance idea of looking at the stars and constructing arguments or relations on this basis. Out of this medium, out of this structure of relations, astronomy just developed. Physics gets its own place outside of the natural sciences only very late . . .

In the nineteenth century.

So pretty late, right? And why? Because at the time groups were interested in the areas that today we call physics—and suddenly one has to take a test, an exam in physics. All at once the interests and activities of specific groups and persons were compressed and refined into their own discipline. My opinion is that here it's down to pure chance. If you have fun with an activity that produces knowledge, it's entirely irrelevant where this activity lies.

The activities might not all be the same—but their names . . .

Ah, what you call them. Without a doubt. Very good.

From the types of roles there's a direct path leading to the surroundings and contexts of such games. And for me there is an important Foerster heuristic connected to these surroundings. Its point of departure is the fundamental proposition that Maturana and Varela put at the beginning of their book, The Tree of Knowledge: *"Everything said is said by someone."*[8] *And your complement* to it—the Foersterian corollary number 1—goes, "Everything said is said to someone."[9] That means that the observer and the reader, the writer and the reader, the speaker and the listener, they form a dynamic unit—for all of them, it's an invitation to dance, to play.

Exactly, that is the idea, that was my supplement to the deep meaning of "Anything said is said by an observer"—"to an observer." Otherwise the reference to observers and speakers loses its meaning, because for me the decisive point is always the being-together, the dynamics.

With the structure of your article, it tries in various ways to get a dynamic going with the reader.

Right, yes. With some articles I only manage to do it for a little while, and simply—a lovely beginning, a lovely ending, a direct connection between beginning and end, and so on. But in one article I made a special effort to make the last sentence identical to the first—and so I stressed in the preface to this work that a reader might begin where they liked; they just had to read to whichever paragraph they started with.[10] I recommended, just out of habit, that they begin with one and finish with twelve.

I'm seeing here, incidentally, an interesting long-term pattern. Those who read your early work from the 1940s will find solutions and proposed solutions—and the problem is sketched briefly at the beginning. You have a scientific problem, work through it, and look for a solution—and the reader "gets something from it" in a very traditional sense. In your later works the solutions that you offer your readers pale in importance to the problem descriptions and the game with the reader themselves. Am I seeing this turning toward the reader, this mutual searching of audience and author, correctly?

You see it very correctly, yes, that became my intention more and more. A problem, if it is a paraphrase of the problem, can already be a solution—and solutions are paraphrases in their turn. If I say "two times two," then "four" is the paraphrase of two times two. Sometimes I manage it, and sometimes I don't manage it, but it gives me joy to put a spin on the presentation of a problem itself and to pull the reader in: "Let's look at it together; we're still moving in the same area, and yet it's all completely different."

Slowly we seem to be striking it rich together, despite your initial skepticism. With our connecting pattern we've reached from the various role-subjects and audience dynamics to activities almost without trying. And a further important element in the Foersterian operator park seems to consist of, if you'll excuse the nominalization, a tracing-back action, which you carry out in the most diverse turns and variations: You incessantly trace nouns back to verbs—things and objects become activities and processes. Because so far we haven't found a more or less suitable label, we'd like—because it has to do with activities—to speak abstractly of a "verb operator" for which various "activity words" can be inserted—"produce," "understand," "create," "build," "generate," "be able to," "do" . . .

"Verb operator," well, this expression isn't especially clear, but maybe it does get us farther! For example, I've noticed the embarrassing consequences and associations result from talking about "knowledge"—I much prefer to use the expression "to understand." If you talk about knowledge, then it's not far to the box in which something must be hidden—something green, blue, earthworms, taxonomies, whatever. "Knowledge" tempts us, almost of its own accord, toward the kind of "nonspeech" that I recently heard from a university president at a graduation ceremony. Universities, he explained, are depositories or warehouses of knowledge that is passed on and handed down from generation to generation. In such statements, almost everything that could go wrong with the imagery is going wrong, from "knowledge" that's "handed down" ("Did you hand down your knowledge yesterday?" "No." "Look out, you'd better do it today!") to "knowledge" that is "stored" like fodder—all these mistakes and wrongs are why students revolted in the 1960s and built the barricades, because they wanted nothing to do with either hand-me-downs or warehouses. The idea of the "Nuremberg funnel" is still haunting us: You pour something into it, shake it all up—and then knowledge is ready. And therefore you'll see again and again how I joke to try to open up these caskets and boxes or to overstep forbidden zones. If someone talks about science (*Wissenschaft*), then I answer, "Ah, you mean the activity that creates knowledge (*Wissen schafft*)." So I'm trying to get this block of ice, science, moving again.

Usually when I bring something like this up with a straight face, people laugh. Why do they laugh? Because I suddenly change an area, like in a joke. Of course these people already knew what "science" was, but all at once they see "science" in a different light; all at once the "iceberg" science (*Wissenschaft*) melts down into an activity that *creates* knowledge (*Wissen schafft*)—and this creation represents a constant activity and does not tempt us to fill up these boxes called "knowledge" with sand, beer, and other stuff. Instead of lecture titles like "Science and Explanation" or "Objective Knowledge," I'd rather have a dialog on the theme "Understanding Understanding." I'm drawn to thinking about understanding as an independent dynamic. I would actually like to be constantly active, and therefore at the center I see *activities*, producing the new, the new, the new.

We've gotten the next operator going so often in the past six days that it's already got to be showing signs of exhaustion; we mean, of course, the "recursion-operator" . . .

You're right; we should let it sleep a bit for the moment. Maybe we'll need it later on.

We'd like to introduce our visit to the next Foerster operator in a little more detail, however, with a Zen koan.[11] Someone goes shopping in the market and says he would like to buy good meat and only good meat. The butcher's answer is, "There is only good meat here." Applied to the area of posing problems, this leads to your operative ability to have only ever offered good and useful problem solutions. The expression "pattern solution"[12] seems to me to be the most suitable, since it indicates an interesting double meaning: on the pattern that creates a connection between fundamental questions and applications, and on the solution that reflects this pattern; "embodies" would probably be a misleading metaphor.

Most of all, I'd like to go into the part of your koan that deals with the "market" and "selling." I don't know whether all of the problem solutions in my intellectual butchers' stall were always of the best quality; I can't judge that. Of quite practical importance, however, was, "Can we sell these problem solutions and results or not?" Problems and their solutions first start to get interesting when you know or find people to whom you can bring these problem solutions closer and to whom you can explain them and who spontaneously react to these with, "Great, we need that, we have to get these results in the fastest way possible."

If you begin to sell problem solutions, then to begin with you are in a really bad position. And I've got to say that I'm still surprised today at how many of my ideas I managed to sell from my market stall over the course of the decades. The persons with whom one collaborates with regard to research funds have, as a rule, very little idea as to your research activities, as to the possibilities and potentials of a field of research or of the perspectives and connections to other problems. But with some ideas I had good luck on the market, for example with the idea of self-organization. "Ha, a fantastic idea!" And so the first of a total of three important early conferences on the theme of self-organization came about, supported and financed by the Office for Naval Research.[13] You see, your question of "good" or "bad" solutions and problems isn't so easily answered. For some of my friends and me, certain holes in our understanding were essential; through these we ventured into some very significant and very decisive problems and resolved, "We want to fill these holes in our understanding." Take, for example, the terminology of cognition and cognitive processes—I believe we were more or less the very first to pose the question of cognition to the world as a research question; since then it haunts every part of the globe. And under the cloak of cognition we began to make language analyses, to examine processes of perception, and so on.

I would like to link up the art of the salesman with the art of the problem-poser.

I'm going to remain a little ironic, however, and hold fast to the problem of selling. What's comical about this market game is that one side has money but doesn't know where it should spend it—and the other side has ideas and problem solutions, but as a rule has a low opinion of the way a market works. Because I knew, however, that the customers like to spend their money where success is guaranteed, I used a special strategy of self-strengthening. I had worked out a complete problem solution; the results lay on the table. Only at this moment did I submit the relevant proposal; described in detail what I wanted to do; and gave very plausible hints as to what results could be expected—they were already lying on the table. They gave me approval for this project—and they received all promised solutions on time, which built my reputation as a successful researcher. As soon as you have a name as a successful researcher you just have to keep submitting what you've just found out—and that is my self-strengthening trick, also known as my magic of time shifting, through which I kept the BCL alive for ten, fifteen years. In later years we became more ambitious and undertook things that we weren't sure would work out—and the problem of marketing became almost insoluble. I even think that it was so easy for me to leave the university at the age of sixty-five because I saw that my planned researches were becoming ever less marketable—I had hit on the limits of resonance.

But I constantly see this phenomenon in scientific work—just take Roger Sperry with the "Split-brain research," which for me represents a very important analytical approach to the workings of the brain.[14] For medical reasons, such as life-threatening epileptic seizures, some patients have the connection between the hemispheres of the brain, the so-called corpus callosum, is severed so that the two hemispheres work totally independent of each other. Out of this emerge very important possibilities for analysis and tests, which we referred to yesterday. But Sperry had the greatest difficulties in getting support from the National Institute of Health, which was financing his research. They introduced ethical arguments, one would not be allowed to conduct such tests, and so forth. And Sperry tried, "These are people who are suffering greatly; I won't cut through Karl Müller's corpus callosum, they are people who are afflicted with dangerous epileptic seizures, and so on." It proved very tiresome to get funding for this very important branch of brain research.

Apropos of "pattern solutions," have you ever worked in a self-organizing manner with an organization, a firm, a university?

With two organizations. One was the BCL. The BCL wasn't so small; there were thirty people working there, and it was part of a big university. It was a system with incomes and expenses. That's one example. The second experiment in self-organization was my first professional activity after the war with the Swedish-Austrian firm Schrack-Ericsson, which had around three hundred or four hundred employees; after the war it was totally destroyed and had to be built up again in order to be able to produce their main products, namely telephones and telephone systems.

The people who worked there, workers, secretaries, mechanics, precision mechanics, engineers, thought highly of me: "This Heinz knows all about the lab, knows how to talk with the director—and talks to us as if he was one of us." When the first unions formed, they immediately elected me to the board of their local union—thus as a socialist Viennese union worker I represented the workforce of the firm Schrack-Ericsson. There for the first time I introduced the principle that every participant in a "managed company" must themselves be managers. This principle was enthusiastically implemented—and everyone made unbelievably constructive contributions to its success. If larger machines were to be purchased, I went to the line managers—because they knew all about this area—and asked what would be required. And then I marched over to Schrack. "For these machines we need this, this and this." Schrack's answer would be "No"—that was clear, he was the boss, and he only had to say no once. But even a boss is capable of learning; he introduced the desired changes in his name. There I learned in an actual company how such a self-organization experiment can make progress.

Such ideas cropped up in the 1980s in the form of "quality circles" and "autonomous working groups." If you were to appear today as a consultant for business or large companies, what would your pattern solution be?

Great question. Yes, I'd have to think about that—so I don't have an immediate answer. I could well imagine, however, that for a single company I would get the ideas of "feedback" and "recursive coupling" anchored within the organization. DuPont is one of the gigantic corporate machines—and two leading managers of DuPont came here and sat right where we're sitting now. The two of them picked my brain pretty thoroughly, and one piece of advice that I gave at the time proved to be constructive and useful. I said to them something like, "Listen, you're constantly stewing in your own juices there; no one knows whether they should believe this one or that one, what someone does or doesn't want to hear, who commands whom in the factory, etc. Attach a separate division with about twenty people to this big system; you could call it the 'industrial research division' if you like. This division should sit 'outside,' so to say, should investigate you and observe what you're doing. And if you listen to the suggestions and ideas this new division develops then an interesting interplay could develop out of that." The reaction of the two DuPont managers was skeptical at first: "We couldn't get a division like that through in the company!" A year later the two managers got in touch again: "We've actually built this new division—and the idea has worked unbelievably well. At first there were some personnel problems, then we searched specifically for some suitable colleagues—and since then the interplay between the outside observation division and the inner life of the large corporation proved to be perfect."

Did you know by any chance what concrete tasks this observation division had to perform?

I think, to a large degree, the taxonomy of system divisions and how one gets away from old routines—here the office employees, there the mechanics, over there the foreman in the engine room, the coach builders, and so on. And slowly they seemed to overcome that so that each competence glided into every other competence, allowing a superadditive composition to emerge, as Gordon Pask would put it, in which work could flow in a creative and mutually supportive way.

That implicitly enables an interesting change of subject. And since we've achieved an initial overview of the first-level operators, we should turn to the probably more difficult question of their parents, the second-level operators. An important operator of this kind, which has come up day after day in our conversations, or rather in your answers, could be called a Foersterian form operator, which in its turn sets in motion first-level operators—inversions, etymologies, and so on.

It's true that I always get up on my form hobbyhorse with every problem. "What is a question's form?" "What form of answer do you want?" "What is the form of this and that?"—the formal aspect of watching and observing. That probably comes from my youthful enthusiasm for geometry. Conception of forms is just very easy for me. As a child I was always irritating my math teachers formally. If I were set an arithmetic problem, I would solve it geometrically. I'd draw, look for an intersection, find the solution. "But no, you're not allowed to solve it like that!" "But why? You wanted the solution; I've drawn it here." This feeling for geometrical forms seems to be an ability to which my parents contributed throughout my life.

I'd like to tell a little anecdote about this. As I've said, geometry was always a lot of fun for me. I liked playing with forms and projections and developed a much better understanding for these than for arithmetic. For my teachers at school, my geometric talents were very unwelcome, since they wanted to teach me algebra and not algebra translated into geometry or universal geometry.

One time a math teacher came to us and did the Pythagorean theorem, $a^2 + b^2 = c^2$. "Here we have a square with sides the length 1; I draw a diagonal; how long is the diagonal?" I raise my hand and say, "Two!" "No, no, my dear Heinz, the Pythagorean theorem teaches us that a^2 and b^2 is the square root of the diagonal, and so the diagonal is not 2 but the square root of 2." "No, no," I answered, "the diagonal is 2, I'll prove it to you! If you take one side, it's 1, and the other side, that's also 1, then we can make some little stairs, yes, half over, half up, half over, half up. How long are these stairs?" To which the teacher said, astounded, "Two." "And now I'll make these stairs smaller, I'll take a fourth, a fourth, a fourth, . . . —two again. Now and eighth, a sixteenth, a thirty-second—and the answer always comes out 2—and the lines get so fine, much finer and thinner than the chalk line we're drawing on the board." "But no, it doesn't work like that." "I've just proved to you that it works like that!" "No, because the Pythagorean theorem . . ." "Please prove the Pythagorean theorem to me!" My math teacher wasn't prepared for that—and so for a short time I was able to keep the diagonal at two. In short, I was a cheeky kid.

At technical college, however, I slid into a deep crisis when I saw that I had only had a very poor grip on algebra. For example, I enrolled in a course on topology with great joy at first, because topology deals exclusively with spatial relations. Topological questions have to do with intersections, projections, neighborhood relations, and so on. So I went to the topology lecture, and there were no drawings to be seen on the board, just one equation after another—"Here is one point, deduced from theorem XY . . ." I was totally desperate: "That's not topology. They're only talking about spheres, discs or other bodies, you don't get to see them." I tried to take an exam, but I bombed it. It was clear to me: "I have to learn algebra, algebraic methods!" And so I withdrew into seclusion for the summer, worked really hard on algebra and on myself—and really drummed algebraic thinking, algebraic thought patterns into myself.

I finally stumbled upon algorithms after I had fallen in love with the *Tractatus Logico-Philosophicus*, which contains truth functions. In these a logical proposition exists as a chain of symbols that can be tested for their truth-values and so on. One might say I had annexed algebra over the summer months—I passed the test in autumn with a respectable grade. Since then I additionally try to project formal thinking into algebraic structures. And even today you can see it over and over, that I see the form of a problem and that I like to represent it in either algebraic or logical form. I thus imagine that I'm able to ride on two different bicycles— one is the formal and the other the algebraic. And then I further imagine that I'm able to concatenate the two bicycles, to chain them together. And if I think back on our trialog, then I notice with astonishment your astonishment that I keep coming back to the form problem.

The concept of "form" is a very ambiguous one. One can read it many ways: For example, in everyday language there is the dichotomy of "form and contents." In Bauhaus design there was the slogan, "Form follows function." In the work of Roger Sperry we find a peculiar inversion, "Function follows form." The concept of form has yet another completely different meaning in the work of Spencer Brown; he doesn't recognize contents, only an inside and an outside. How do you yourself cope with the ambiguities of the form concept?

I am always seeing form as a structure of relations that can also have different structures of relations than their elements. If someone asks me, "What is knowledge?" then I ask back, "In which form would you like to see an answer? I can give you a dictionary answer. Let's look it up together in the dictionary and see what progressive things it has to say about "knowledge." Ah, here it is: knowledge—blah blah blah. Is this answer satisfactory?" "No," replies the other, "absolutely not. This dictionary answer seems to me to be totally inappropriate to the question of knowledge." "Well, what may I offer you as an alternative? Would you like to have an etymological answer to what 'knowledge' is? The German for 'knowledge' or 'insight,' *Erkenntnis*, comes from *er-* and *Kenntnis*, from *kennen*, to know. And *er-* is a prefix that works very like the other German prefix *be-*. *Be-* and *er-* possess great similarities, so one could translate *er-Kenntnis* into *be-Kenntnis*, meaning "confession" or "declaration," or one could replace *kennen* with another word for "to know," *wissen*, giving us *er-wissen*, "to make sure". So *Erkenntnis*, knowledge, goes through a slow process of *er-wissen*, of making sure, which brings me to seeing something new, understanding something new. With this, we'd have knowledge (*Erkenntnis*), making sure (*er-wissen*), confession (*Bekennen*), and consciousness (*Be-wissen*) all brought together into a nice configuration." And with that we've got an initial form that shows how the concepts around *kennen* are arranged, which symmetries exist, which semantic triangles and rectangles are opened up, at which points it would be possible to cut through, and so on. I mean all of that with a problem of form.

Now we come to another second-level operator, which switched itself on right at the beginning of our conversations, probably not by chance. I would like to call this operator "here and now," ambiguously speaking. This operator transforms the apparently distant past or far future within a conversation—"here and now."

I'd like to tell a personal story about that. In my childhood, from when I was three till I was seven or eight, I grew up in my mother's family, with my maternal grandmother. Why? My father was drafted in the first weeks of World War I and was quickly captured by the Serbians, who held him for three years until he was exchanged in 1917. And so for a long time I grew up in my grandmother's house and with my mother's family. My grandmother was an extraordinary woman, one of the first advocates of women's rights, an early feminist if you like. She published along with others the *Dokumente der Frau* (Women's Documents) and led a salon in which likeminded people met.

My grandmother had a great influence on me through her sayings. As a child one must constantly overcome terrible disappointments—you lose something, you break a little toy, you cry, you're sad. "Now why are you crying?" "I just broke this little knight. The horse's feet broke off . . ." "Listen to me," my grandmother would console me, "everything is here and now. You think there used to be a horse. But the horse is here and now, it is here now—there is nothing that has been." This "Everything is here and now" became a mantra for me.

Interestingly, your reaction to the "here and now" came from the "there and then." Your anecdote—and with it many other stories that we carry around with us, our own and others'—stands in prima facie opposition to an operator that transfers everything to the "here and now."

We have only just come into being, here and now, with all the traditions, whether they've come about genetically, personally, historically, or any other way. They likewise belong to our here- and nowness; you can listen to them if you like. You can listen to them, and of course you can also refuse to listen to them—then you exist in a kind of vacuum, which does, by the way, happen to many people, unfortunately.

For me, another operative area that is very closely coupled with the "here and now" is that you generally try to choose your pattern solutions and pattern examples so that on the one side the individual as agent moves in the center, but that on the side the responsibility of the individual for themselves but also for the condition of their environment is strongly emphasized.

I think I picked up this attitude early on from my grandmother's circle. At the time one was very conscious of a responsibility with respect to society and developed a fine feeling for questions such as, "What is necessary right now? What is urgently needed right now? What is a strong, unfulfilled desire?" If you yourself felt, "It would be nice if people had this or that at their disposal," then from this wish emerged an action, an explicit or implicit program that created knowledge toward this aim. I'd like to tell about an experience from which I learned very early on to take responsibility for finding the solutions to problems myself. When I was a little boy we would often spend the summers in Salzkammergut. One afternoon a thunderstorm was on its way; the swallows were flying very low, and my parents called to me, "Look, bad weather's coming, the swallows are flying so low." I asked back, "Yeah, why do the swallows fly so low when bad weather's coming?" And my parents said, "Because the mosquitoes, the flies, the insects and the gnats, they all fly very low when bad weather's coming." Then I cheekily asked, "But why do the mosquitoes and the insects fly so low when bad weather's coming?" Bam, and I got my ears boxed. Well, then I knew, that seems to be a very fundamental question, which one can't answer. And so at the time I drew from this the following conclusion: If you want to have the fundamental questions answered, you've got to look after it yourself.

Back to responsibility. I've noticed in many cases that my diverse research programs spring from an implicit wish to reduce a general flaw, most often a social flaw, in one form or another. Therefore, I draft the question so that in a certain sense the attempt to eliminate or relieve this flaw is contained within it, so that this social responsibility is addressed. If I take this idea seriously, that one fits oneself into this cycle of seeing and being seen in a responsible way, then it becomes more and more clear that I cannot delegate decisions about myself and responsibility for myself. My actions force me to take the full responsibility for them as well. It is seeing, not portraying; creating, not obeying; freedom, not force. And this point—if you like, this responsibility operation—showed me that an ethics must be "implicit," must "show" itself. I first mentioned this in my Paris article on "Ethics and Second-Order Cybernetics."[15]

For the work on this article I naturally referred to Wittgenstein, especially to the passage in *Tractatus*: "It is clear that ethics cannot be voiced." Well, that gives everyone a fright: This Wittgenstein, he doesn't want to talk about ethics at all, because ethics cannot be voiced. I've seen lots of people who have been frightened when I've quoted this proposition. Some see it as virtually laying a foundation for something wicked. But for me it's not wicked at all because Wittgenstein is warning that if I begin to put ethics into words, I will be moralizing, well, and then ethics becomes a moral sermon, and I have to avoid that above all else. If I'm preaching morals, I'm always saying to others: You must do this, or you may do that, or you may not do that, and so on. Ethics on the other hand doesn't refer to the other but to one's self. I must do this, I should do that, and so forth, and so, I would like to conclude with a historical comparison.

The great sky magicians, Albertus Magus for example, made it clear again and again that the astrological idea—that the stars influence people—is a completely false interpretation. And Albertus Magus invited a different perspective: The world is like this, and from out of this world there arises another—and we are all sitting in on this development together. So I am just as responsible for what Jupiter does—as Jupiter is, on its part, for how I act. Nothing is influenced from only one direction. We are all in a room, in space, in this world-space, in a thought-space where one thing is connected to the other over and over again.

A very important point, which is also already our fourth second-level operator, is represented by second-order operations. You very much like to build the form of your problems into a second-order problem—and you form them so that what you're talking about becomes part of the processing and, so to say, contains itself.

You've picked up on that well. I can still remember it so well, when the Macy Foundation invited me to write a preface for a cybernetics conference. With great delight I wrote a preface in which I raved on about this unbelievable new geometry, about the new form of circular arguments. I referred to the transition of linearity to two-dimensionality and worked myself up to a *salto mortale*, to a formal leap of death. And it actually was a *salto mortale* because they explained to me shortly and succinctly that they didn't need that kind of thing. They wanted a nice story about the Maxwell regulator, the water closet and the thermostat, which didn't interest me in any form whatsoever. What thrilled and fascinated me, contents-wise, were the questions about the new logical relations that arise from circular causality. Perhaps I should emphasize why I hold these second-order concepts to be so important: As soon as you withdraw to the area of the second-order—the understanding of understanding, the knowledge of knowledge—the first-order problems are suddenly illuminated in ways that you can't perceive on the first level.

I could give many examples now, but I'll limit myself to one that gave me many surprises. In cybernetics, the word "purpose" comes up very often—and on the first level, one will slave away at describing the purpose of X, Y, or Z. On the second level, however, the question is, "What is the purpose of purpose?" As soon as I had this question before me, I was led to the following considerations. If I postulate a purpose that I'm aiming toward—Aristotelian *"causa finalis"*—then I don't have to postulate step-by-step the transitions from the condition in which I now find myself to a condition that I'm aiming at, instead I can already determine the final aim, the *"causa finalis,"* in advance. The purpose of purpose is the determination of goals without having to consider the ways, routes, and trajectories right away. And that also hits on the point that Norbert Wiener and the early cyberneticists recognized very decisively: What does the navigator who wants to get into the harbor do? In a purely physical way I can never determine what course the ship will take to get into the harbor. The wind is roaring from the left or right, obstacles suddenly tower before you, other ships are crossing. Laying it out like this just isn't possible. But as a rule the navigator manages to get the ship safely to harbor because he constantly sees the divergences from the course and can correct the steering—and in this way finally lands in the harbor. And that's why the introduction of purpose makes cybernetic and physical sense: It's purposeful. It relieves me of the burden of having to constantly deal with the next step.

We've now marched many Foerster operators backward and forward in review. Have we neglected a particular point in our conversations that seems particularly important to you and which we absolutely must mention in our circle of Foerster modules?

That would be all my love affairs with my environment. It's probably always the case that I enter into loving relationships with others and therefore try to dance with others. And because I probably wouldn't want to dance with a stranger, with someone I didn't like, I always immediately see my partners as loveable and dance-loving people. That astonishes a lot of people. I go into a store and see someone who's crying. I ask, "Ah, what's the matter?" "My grandmother has died." I would try to comfort them.

If you walk through Pescadero, it's a succession of social dances.

Well, it's very funny, isn't it? I address the inhabitants of this little village as *menschen*, and therefore I'm a *mensch*, them as well, not just a customer or someone who is buying gas or looking for stamps. I think that's what it's all about.

Toward the end of our conversation, let's start a counterfactual game. Imagine that you were twenty-five years younger and still had a laboratory with highly motivated people to play with you. What problems would you want to work on today with the new technological possibilities and with the current state of knowledge?

Great question, but you'll be disappointed that my answer is rather dull. If someone offered me $500,000 now, if someone said to me, "Heinz, you can invite six people of your choosing, you've got the laboratory of your dreams"—what would I do?

You would also be twenty-five years younger.

Well, that helps even less, I'm actually very glad that I'm twenty-five years older. But above all I'd like to undertake something that seems fruitful to me, and that is *to take recursivity seriously.* Unfortunately or perhaps ironically all of the recursion topics have slid down into what today is called chaos theory, fractals, into all the wonderful magic shows, which one can sell graphically, numerically, and verbally to the *New York Times* with lots of incredible catchwords and phrases. I would take this area more seriously, take it up a level or even whole stories; that would be my great research aim. Unfortunately the analysis would stay exclusively on the level of numbers for the time being, either on linear level or on the level of complex numbers. My feeling is that one could take these recursion mathematics into totally different areas, maybe in linguistics, semantics, fields of action, and so on and could examine under which conditions viable stabilities develop—nonstabilities remain invisible, they disappear. My central question could be: What are the visible forms into which dynamic systems can "slip" or "drift"? The first signs, which we know from the field of numbers and also complex numbers already indicate how such systems operate in other fields. That would be my research program today, and for it I would group linguists, biologists, and of course mathematicians around me. My feelings tell me that one could bring to light many, many more, infinitely more important insights, especially for our social problems.

In what regard would your research program differ from the writings of Luhmann, who, as a single author, has already brought out an imposing library on autopoiesis, recursivity, and the second order?[16]

First of all, I don't *know* what Luhmann has written thus far; I couldn't pass a test with him asking me, "Explain briefly the essential points in *The Economy of Society* and *The Science of Society*." I've met Luhmann several times, and we talked—he was in Pescadero, I in Bielefeld—and when I read one of his publications, it seemed so long-winded to me that it was really hard for me to get to the end. My feeling is that Luhmann can only work in the forms that are close to him as a sociologist and that are based on the concepts introduced by classic sociology.

Perhaps the form, the Luhmann answer-form for the question "What is society?" is necessary, but it's not sufficient to touch on the depths of the problems that arise in these contexts and that have to be dealt with. When I gave my lecture in Bielefeld that time, I said, "You see, the problem that sociologists should actually deal with is the 'Vance-Owen problem'!" What is this problem? Vance and Owen, they're the two people who tried to initiate a conversation between the Serbs and the

Croats when the conflicts broke out in the 1990s. It's clear to see that they brought about no such conversation. No one listened to them; they spoke in languages that neither side used. Here are the problems that today's sociologists have to deal with—they are nontraditional and of a different quality than those that sociology has normally considered as its domain. And if someone asks me, "Say, Heinz, how would you differ from Luhmann?" then I can't answer the question because I don't know how and what Niklas Luhmann analyses. But when I listen to Luhmann, when he reads his papers and when I flip through his fat books, I see that in many instances my suggestions about studying "closure," "recursion," and so forth haven't been taken up, that such ideas would be applicable in the areas where I had hoped they could be applied.

The example of family therapy—would that be a part of such a program? Or take the idea of self-organization in the field of management—might this be a possible indication of the direction you would take?

Oh yes, absolutely. For example the management styles, there are so many of the classic "management axioms" that are counterproductive, and right at the start I would say, "Look, let's get rid of all these jokes, let's forget the 'bottom line,' let's start with the 'top line'!" Let's scrap all these self-evidences and then see what's left. And then we'll come across the madness of corporate downsizing. Because no one has the courage to talk about throwing out and dismissing anymore, you use the expression "downsizing." Here again I would start out, "What a fabulous idea, of course we can earn more if we downsize the company; why don't we shrink it down more and more till there's no one left at all; then we'll all be trillionaires." One quickly sees that these concepts are so crazy that one hardly knows how one should even start talking to such managerial acrobats. But then it happens again and again, that I'll lay out a couple of unpleasant subjects before managers, whereupon some will abruptly break the contact, but some urge me, "Tell us more, Heinz, keep going!" And then I talk about the important McCulloch theorems because I think that they're formulated in such a wonderfully accessible way that maybe managers could make something of them and start managing in a similar manner. Or take, for example, my fundamental theorem of management: In a company, everyone has to be able to take on the role of a manager.

Let's assume that everyone in a company is a potential manager and then let's look at what new relations that has to result in. Then the new question for management is "Why do people like to work with me so much?" and not "How do I motivate the people under me?" You should never have to say, "I work for someone," but rather, "I work with someone." If people are doing something with me, then each of them can explain what is to be done just as well as or in some cases better than I can. I'm ready to listen, which means I'm expecting a better solution than any I can offer. As soon as you develop this kind of attitude, a company begins to function completely differently; then incredible reactions can be let loose. People are enjoying themselves: "I'll show that Heinz, we're going to get better." And the lethargy and apathy vanish. It really doesn't matter what happens to me. Such attitudes can be taken advantage of by others under certain circumstances, but these ideas about management and self-organization seem to be something of an algorithm for my activities. In the laboratory, too, I didn't need to appear as the author on a paper as long as the group stayed lively.

If there's no sufficient necessity of following Luhmann's trail, how about the path laid by Stafford Beer in the field of organization, of firms, but also in the area of national political consultation? There's a very interesting series of books by Stafford Beer.[17]

Stafford wrote absolutely fabulous books in his time!

Would suitable signposts have been set up here?

Well, sadly Stafford uses several metaphors that proved to be misleading, and nevertheless he still maintains his unfortunate association with them. It would be hard to wean him from the discrete charms of his metaphors. I got to know Stafford Beer in 1958 in Namur. He came to the conference in fine English tweeds—Stafford was already tremendously rich because he was an industrial consultant. We, on the other hand, were wearing out our last pairs of shoes; Ross Ashby hardly owned a wearable jacket. Our first meeting, however, turned out to be really funny and amusing; apart from Ross and myself, Gordon Pask was also present. Stafford gave us all an invitation—and so we went to this beautiful restaurant in Namur's castle or citadel, an exquisite, expensive restaurant in which Stafford ordered champagne and caviar.

Be that as it may: At this time Stafford's great task was to modernize the whole of English heavy industry; he had to get the ancient hunks of junk they had standing around going again operatively. I'd like to give a quick sketch of Beer's solution to this problem so that you'll be able to see what a brilliant thinker and organizer this Stafford was. English heavy industry got into difficulties after the Second World War because the losing states built their destroyed industries anew while the victorious Great Britain kept producing in its old factories. The great auto industry constantly needed new pieces of metal for its bodywork. But the metal had to be a very specific thickness because if it was too thick or too thin then it would have a detrimental effect on the bodywork—and an auto manufacturer had to scrap such bodywork immediately. Well, English heavy industry wasn't managing to press metal to the required thickness. Their rolling mills weren't precise enough for this purpose—the machines were already so outdated and had so much clearance that they could no longer be fitted and adjusted exactly. English heavy industry lost large orders because of this—and at this moment Stafford Beer came onto the scene: "Listen, all we have to change is that we have to measure our metal output precisely. Measure how thick the metal coming out is, and then we sort the metal into piles according to their thickness, so one pile with 0.9mm, one with 0.95mm, one with 0.98mm, and so on." This way the outdated steelworks worked more precisely than any of their competitors could have. Even modern machines had too much clearance and fuzziness to cut precisely to a hundredth of a millimeter. On the back of Beer's idea, English industry quickly started to look up again; soon it was once again more precise and reliable than its competition.

Let me ask once again: In Stafford Beer do we find informative starting points for recursivity analysis that are according to your tastes?

I would, unlike Stafford, situate the self-organization process on a much lower level and would more clearly emphasize the differences between the organization of a nervous system and of a society. I would like to tell you the story of our Chilean adventures quickly because it makes both these points clearer. I spent several months in Chile with Stafford to support Salvador Allende's economic experiment. When Allende managed to get into government, he took a very dear young friend of mine into the government, Fernando Flores, a brilliant man who developed fabulous new economic and social ideas.

Allende named Fernando minister for the entire economy and gave him a clear assignment: "You're building a new form of social economy in Chile!" Fernando—he was twenty-six years old at the time—called on his great master, Stafford Beer, and managed to invite him to Santiago to redesign the Chilean economy. Now, Fernando Flores also happened to be a good friend of Humberto Maturana and asked him to collaborate with them as well: "Stafford Beer is coming to us, the one who came up with the great metaphor of looking at companies, work, the whole economy as a nervous system."

It so happens that at the same time a guy from the University of Illinois, Heinz, along with his wife, Mai, had come down to visit Humberto Maturana. And so in Santiago we celebrated an unexpected reunion. Now that it is legitimate to see an economy as a nervous system, the question is just: Does this metaphor hold water? Stafford was convinced that an economy had to be organized like a nervous system. An economy has to operate like a nervous system, it has this many functions, the individual functions have to be handled like this, etc. The first thing that he introduced was a so-called observation room: In order to know what we're producing, we have to form a connection with all the sites of production in Chile. Thus we have to know what the auto manufacturers are producing, what the tire factories are producing, what the vegetable farmers are growing, and so forth. So we sat in this observatory in Santiago where all the data from the different companies was coming in and where one could give orders that in the future more potatoes, more tires, more motors, more oil should be produced.

Now, this room was not called the "control room," the "steering center," or the "cerebrum"; instead, it had the innocuous name "observation room." This move turned out to be political gold, because if one had called it the "control center," then the criticism would definitely have come: "Ah, you're the 'great dictators'; we've got to dance to your tune!" Then we could pull back, "No, no, we're just here to observe." Soon we had to conclude that you cannot project a nervous system onto an economy. You can't transfer an economy onto a system that incessantly organizes itself.

In the Staffordian economy there was too much preorganization, which was fixed in his eyes. It lacked the freedom to allow dynamics to develop that were no longer controlled by this strict model. The control was placed too high up and not where it should have been, namely down with the companies and the individuals. The experiment was dramatically ended from outside. Allende was murdered; Pinochet seized power; Fernando Flores, who had brought Stafford Beer to Chile, was finally arrested and held prisoner for a long time in this horrible prison camp on an island in Tierra del Fuego.[18]

If one now—we're moving toward the end of the sixth day—looks at the development of these Foerster operators over time, if one tries to find an arrangement for these first- and second-level operators,[19] then a peculiar phenomenon emerges: Around the time of your retirement, in the second half of the 1970s, then your approaches, your heuristics, these "Heinz modules" seem to reach such a pitch that they enter into a special set of eigenvalues. My impression is that over the decades you've worked on a series of paradigmatic cases that you have condensed and compressed—until finally an area was reached in which operatively these heuristics created a special set of eigenresults.

Aha, it's very interesting that you've observed that. If it happened that way, then I didn't do it on purpose—that's the way the ball bounces, isn't it? My retirement freed me all at once from a pressure, the pressure to keep a biology lab going, to work with people constantly, and so forth. And if the pressure from within is relieved, then there's the pull from without—people in Paris wanted a lecture from me, invited me, people in Hamburg wanted to hear me "live." In most cases people gave me a free choice of topics—and so to the people of Hamburg I suggested that the understanding of language is of importance to psychiatry—language is the only medicine in their possession, all they can do is talk with people. Therein lies the magic of family therapy, that it doesn't prescribe medication; rather, they talk with people—that's why language becomes the therapeutic medium. At the world congress of social psychiatry I absolutely wanted to talk about language—thus the strange title "Language of Magic." Sometimes, however, someone would stipulate a title, but that is often very agreeable for me, too. If people dictate a title to me, then when I talk about it I know that I'm complying with their wishes.

Whether I satisfy them in the sense that they've anticipated, that, obviously, I can't judge. All I can say is, "If a theme was put before me, how would I deal with it?" That way I've always got the excuse that for the organizers the most important thing was that Heinz make an appearance, otherwise they would have probably booked Fritz or Max or Emil.

In Wittgenstein's Culture and Value, *we find the sentence: "Thoughts at peace. That is the goal someone who philosophizes longs for."*[20] *Could one say, in a variation, Those who strive recursively cannot provide, but may lead to thoughts at peace?*

If I want to achieve "thoughts at peace," then I'll run through recursive operators toward a stability—peace runs through the recursivity, it runs back through, back through, till it finally stabilizes itself. But this kind of peace can only ever be short-lived and can stretch itself over some thoughts, but not all of them.

Heinz, we're drawing slowly to the end of our trialogs. Actually, it would be an interesting idea to compose a book on your heuristics and points of view in the style of a conversation.

That would be an interesting book, yes. I would buy it at once.

We'd have an especially marketable idea. Over six long days, we—together with you—invent and program a Foersterian thought machine. The question then is just: "How would one do such a thing?"

My point is this—it is fundamentally impossible.

SEVENTH DAY

Rest, Rest, Rest, Rest

Now, you may say, "O.K., so what?" But, ladies and gentlemen, if you say, "so what?" to anything, you will not see anything.
—HEINZ VON FOERSTER, *Understanding Understanding*

We are satisfied that the earth is round.
—LUDWIG WITTGENSTEIN, *On Certainty*

Thus the heauens and the earth were finished, and all the hoste of them. And on the seuenth day God ended his worke, which hee had made: And he rested on the seuenth day from all his worke, which he had made.
—GENESIS 2:1-2

HEINZ VON FOERSTER CLOCK[1]

XII

The environment is experienced
as the residence of objects,
stationary, in motion, or changing.

XI The environment contains no
information; the environment is
as it is.

The logical properties of **I**
"invariance" and "change" are
those of representations. If this
is ignored, paradoxes arise.

X The information associated with
a description depends on an
observer's ability to draw
inferences from this description.

Formalize representations **II**
R, S, regarding two sets
of variables {x} and {t},
tentatively called "entities"
and "instants" respectively.

IX The logical structure of
descriptions arises from the
logical structure of movements.

Contemplate relations, **III**
"Rel," between representations,
R, and S.

VIII Terminal representations
(descriptions) made by an
organism are manifest in
its movements.

Objects and events are not **IV**
primitive experiences. Objects and
events are representations of relations.

VII A formalism necessary and
sufficient for a theory of
communication must not contain
primary symbols representing
communicabilia (e.g., symbols,
words, messages, etc.).

Operationally, the **V**
computation of a
specific relation is
a representation of
this relation.

A living organism is a third-order relator
that computes the relations that maintain
the organism's integrity.
VI

EPILOGUE IN HEAVEN

> The problem is not truth, he answered, the problem is trust.
> —HEINZ VON FOERSTER, *Understanding Understanding*

> If someone believes that he has flown from America to England
> in the last few days, then, I believe, he cannot be making a
> *mistake.*
> —LUDWIG WITTGENSTEIN, *On Certainty*

> Therefore the Lord God sent him foorth from the garden of
> Eden, to till the ground, from whence he was taken.
> —GENESIS 3:23

On the way back from San Francisco to New York we're sitting—"tired and happy" would be a fitting phrase—in a flying and, for the next hours, hopefully, totally trivial machine. In our hand luggage we are carrying, in the form of sixteen cassette tapes, the extracts from one week of *Hissing, Grunting, and Rattling Sounds From Rattlesnake Hill*, also known as *Dialog from This Side of Eden* (since the Foersters' estate is on West Eden Road). Heinz von Foerster made a special point when—in a coda to the sixth day—he added regarding our great design plan for the Foerster machine:

Since such a book cannot be made; everything that you produce here is your invention—and therefore it is your responsibility to invent a

Foerster that you would like to have invented and that emerges from your form of "finding." Therein lies the Müllerean appeal—and I'll be fascinated to get to know him. About the Heinzean Heinz, the one which I am myself, I know nothing or much too little—so I need the Müllerean Heinz to get to know the Heinzean Heinz better.

And since according to one of the two fundamental principles of Foersterian pedagogy one can learn even from the dumbest, this part of the "Know Thyself Better" game was set for the time being. The ways and means of the Foerster creation *sub specie* Müller and Müller was, in a certain sense, left open to us. If something seems to be fundamentally unattainable, every attempt comes equally close to the mark.

Beneath us the landscape of the Rocky Mountains unfolds: white mountain ranges with dark flecks, white and black.

I can tell you a little story that I really like and that touches on the problem of explanation. My son Johannes was a volunteer in Africa with the Peace Corps—and he made a very dear friend there, a Nigerian named Ignatius. After the tragic death of our son, we invited Ignatius to study at the University of Illinois and to live with us.

Ignatius came then in September, straight from a small village. I soon noticed that he was interested in photography. On the table in his room he had a picture of the chief of his village, a second of his mother, and so on. For Christmas I gave him a camera. And he immediately started taking photograph after photograph with it. After a couple of days he picked up the developed film, came to me and said, "Heinz, the photo lab has lied to me and cheated me. I used color film to take a photo of my village chief and my mother on my table—and now the pictures are black and white." I answered him, "Your original is also black and white, so you can't get anything other than a black and white picture." "No, no," he said, "I've bought color film, now the colors have to be in the pictures as well." I tried to explain the problem once more, but he just stuck with it: "Heinz, they've cheated me because it was a color film—and these pictures are just black and white."

Slowly I saw that I was having difficulties. Thank God my very intelligent son Thomas came to visit. He had just started vacation, and I said to him, "Tommy, you teach physics. Here is my friend Ignatius, tell him that you're only ever going to get a black and white picture if you photograph a black and white picture!" Thomas made extraordinary efforts, but Ignatius still felt cheated.

Then a young man came along, John White, who had taught in Africa himself. I asked him as well to explain the black and white problem. White went to him and just said, "It doesn't work!" And Ignatius understood. I find this point central to the question "What is an explanation?" All the efforts that Tommy and I were making were totally in vain: "It doesn't work!" That explains everything; we didn't need to keep racking our brains.

"One must throw away one's own standards of explanation and climb down from the ladder that one has climbed up." That would probably also be a fitting variation on the theme of "explanation." Interestingly, in the coda to the sixth day, a story about Victor Frankl came up that suggests the opposite alternative. As Heinz von Foerster told it, shortly after being liberated from imprisonment in the concentration camps, a man lost his wife and sank into a terrible weariness with life. During a therapeutic session, Victor Frankl confronted him with the fictional possibility of creating a person exactly like his deceased wife and asked whether he would actually want this. The widower declined, however, and after a short hesitation, through this dialog found his way back into the "sensible world."

I asked him, "How is that possible, what happened there?" And then Victor Frankl smiled in his very special way and replied, "It is very simple. We see ourselves through the eyes of others. When his wife died, he was blind, but once he had seen that he was blind then he could see again." For me this metaphor is still totally incredible, incredible.

One must raise up one's own standards of explanation and find a ladder that one can climb up—in this form a further variation on the metamagical theme of explicability almost suggests itself. Up a ladder, down a ladder, away from a ladder, ladder gone—for such decisions a ladder daemon would certainly come in handy, one which had two basic operations, with one sign meaning "up" and one meaning "down."

The Foersterian answer to this daemonic offer would probably not be much different from the reply to the offer in our first conversation about a daemon on the house, a "Laster daemon" (Laplace + Foerster) that would be able to differentiate between trivial and nontrivial systems.

That is a very lovely daemon—and yet I still would not get together with it. My daemon functions completely differently. The beauty of explanations is that you can get them to run recursively.

And of course Heinz would have added immediately that explanations—like tastes—have to be different.

What do explanations do: explanations connect two descriptions semantically. How this happens in individual people is different because the semantic structures in people are different. What is a very exciting question for one person won't interest the other for long.

Meanwhile we had flown over the Continental Divide, and beneath us the world started to orient itself toward the east. For the next weeks and months *one framework* would determine our activities—thus far the coming task was certain: the transformation and metamorphosis of our long "trialogs" on tape into the format of a dialog with the reader. Toward what results will we drift or slip with our "magic of transubstantiation"? At least the maximum or the optimum can be set down in quotation form, this time from the Wittgenstein fund: "What we are supplying are really remarks on the natural history of human beings; we are not contributing curiosities, however, but observations that no one has doubted, but that have escaped remark only because they are always before our eyes." With some turns and inversions we could get from this the minimum expectations for the text passage that is to be created: *What we are supplying are really remarks on the cultural history of human beings; curiosities and observations that everyone has already noticed and that no one has doubted because they are constantly happening before our eyes.*

We have definitely left the mountainous region of the Rocky Mountains and we find ourselves over the plains of the Midwest. Observed from above the difference is only marginal; snow transforms even the flatlands into a kaleidoscope of mostly white areas with a few dark patches. In the coming time we would spend so many hours, days and weeks pouring these trialogs into a form both readable and worth reading— even if we were still far from that during our journey over the American continent. What we most needed for it was a very special transformation artist: *What we need are transformations of observations on natural and cultural history, which everyone notices and no one doubts, into statements free from doubt, which only escape remark because they are constantly before our eyes.* And as has already happened before during the six days, the surprising Heinz von Foerster pushes himself between us to make a point on the final (re)presentation (description):

> That's a very lovely ending for a book—and yet I wouldn't go for it. My book endings function completely differently. The beauty of book endings is that you can get them running recursively, round and round.

NOTES

FRONTMATTER

1. Hans Hahn, "Superfluous Entities, or Occam's Razor" (1930), in *Empiricism, Logic, and Mathematics: Philosophical Papers*, ed. Brian McGuinness (Dordrecht: Reidel, 1980), 4.

2. See "Im Goldenen Hecht. Über Konstruktivismus und Geschichte. Ein Gespräch zwischen Heinz von Foerster, Albert Müller und Karl H. Müller," *Österreichische Zeitschrift für Geschichtswissenschaften* 8 (1997): 129–143, esp. 136.

3. See Ludwig Wittgenstein, PI § 510: "Try to do the following: *say* 'It's cold here,' and *mean* 'It's warm here.' Can you do it? And what are you doing as you do it? And is there only one way of doing it?" Wittgenstein rejected the idea of private language.

FIRST DAY: BUILDING BLOCKS, OBSERVERS, EMERGENCE, TRIVIAL MACHINES

1. "The environment contains no information. The environment is as it is." Heinz von Foerster, "Thoughts and Notes on Cognition," in *Cognition: A Multiple View*, ed. Paul L. Garvin (New York: Spartan Books, 1970), 25–48, here 47; UU 189.

2. On Marie Lang, see Heinz von Foerster, UU 325 ff.

3. *Rashomon*, directed by Akira Kurosawa, 1950.

4. See Heinz von Foerster, UU, 293.

5. See George Spencer Brown, *Laws of Form* (New York: Dutton, 1979). The motto is placed before the text starting with page 1.

6. See Karl R. Popper, *Unended Quest: An Intellectual Autobiography* (New York: Routledge, 1982), 216: "For example, men may have invented the natural numbers. . . . But the existence of prime numbers . . . is something we *discover*."

7. See Murray Gell Mann, *The Quark and the Jaguar: Adventures in the Simple and the Complex* (New York: Freeman, 1994).

8. Voltaire, *Candide ou l'optimisme* (Geneva, 1759).

9. As a résumé, see Ilya Prigogine and Grégoire Nicolis, *Exploring Complexity: An Introduction* (New York: Freeman, 1989); Ilya Prigogine and Isabelle Stengers, *Order Out of Chaos: Man's New Dialogue with Nature* (New York: Bantam Books, 1984).

10. Leibniz, *Theodizee* III.

11. See John D. Barrow and Frank J. Tipler, *The Anthropic Cosmological Principle* (Oxford: Clarendon, 1986).

12. The eminent Austrian dancer Grete Wiesenthal (1885–1970) was an aunt of Heinz von Foerster. See UU 325 ff.

13. See John Archibald Wheeler, "Information, Physics, Quantum: The Search for Links," in *Complexity, Entropy, and the Physics of Information*, ed. Wojciech Herbert Zurek (Redwood City, Calif.: Addison-Wesley, 1990), 3–28.

14. Ludwig Wittgenstein, PI § 24.

15. Karl R. Popper, *The Poverty of Historicism* (Boston: Beacon Press, 1957), introduction.

16. Warren S. McCulloch and Walter H. Pitts, "A Logical Calculus of the Ideas Immanent in Nervous Activity," *Bulletin of Mathematical Biophysics* 5 (1943): 115–133.

17. John von Neumann, *The Computer and the Brain* (New Haven, Conn.: Yale University Press, 1958).

18. See Heinz von Foerster, *Through the Eyes of the Other*, in *Research and Reflexivity*, ed. Frederick Steier (London: Sage, 1991), 63–75.

19. Prigogine and Stengers, *Order Out of Chaos*.

SECOND DAY: INNOVATION, LIFE, ORDER, THERMODYNAMICS

1. Ludwig von Bertalanffy, *Theoretische Biologie* (Berlin: Borntraeger, 1932, 1942).

2. See Grégoire Nicolis and Ilya Prigogine, *Self-Organization in Nonequilibrium Systems: From Dissipative Structure to Order through Fluctuations* (New York: Wiley, 1977).

3. See J. L. Locher, ed., *The World of M. C. Escher* (New York: New American Library, 1974).

4. *Go West*, directed by Edward Buzzell, 1940.

5. See Heinz von Foerster, "On Self-Organizing Systems and Their Environments," in *Self-Organizing Systems*, ed. Marshall C. Yovits and Scott Cameron (London: Pergamon, 1960), 31–50.

6. The Biological Computer Laboratory was directed by Heinz von Foerster at the University of Illinois, Urbana. The reputation of the BCL is legendary today because of its transdisciplinary praxis of research and teaching. Important members of the BCL besides Heinz von Foerster have been W. Ross Ashby, Herbert Brün, Gotthard Günther, Lars Löfgren, Humberto Maturana, Gordon Pask, Alfred Inselberg, and Paul Weston. The BCL was closed following Heinz von Foerster's retirement. The scientific work of the BCL is well documented by a microfiche edition. Of great interest is the liberal approach to teaching and learning, with systemic commitment to involving students. *CoC* is an excellent document of these educational principles. See also Albert Müller and Karl H. Müller, eds., *An Unfinished Revolution: Heinz von Foerster and the Biological Computer Laboratory* (Vienna: Echoraum, 2007).

7. Lars Löfgren, "Recognition of Order and Evolutionary Systems," in *Computer and Information Sciences*, ed. J. Tou (New York: Academic Press, 1968), 2:165–175. Reprinted in *CoC*.

8. Regarding the Turing Machine, see Alan Turing, *The Essential Turing*, ed. B. Jack Copeland (Oxford: Clarendon Press, 2004).

9. Warren S. McCulloch and Walter H. Pitts, "A Logical Calculus of the Ideas Immanent in Nervous Activity," *Bulletin of Mathematical Biophysics* 5 (1943): 115–133.

10. See John von Neumann, *The Computer and the Brain*, (New Haven, Conn.: Yale University Press, 1958).

11. Löfgren, "Recognition of Order and Evolutionary Systems."

12. Heinz von Foerster, "Notes on an Epistemology for Living Things," UU, 251.

13. See Heinz von Foerster, "Disorder/Order: Discovery or Invention," in *Disorder and Order: Proceedings of the Stanford International Symposium*, ed. Paisley Livingston (Saratoga, Calif.: Anima Libri, 1984), 177–189; UU, 273 ff.

14. Erwin Schrödinger, *What Is Life?* (Cambridge: Cambridge University Press, 1944).

15. Foerster, "On Self-Organizing Systems and Their Environments."

16. Foerster, "Notes on an Epistemology for Living Things."

17. Ibid., 117.

18. Humberto Maturana, *Biology of Cognition* (Urbana, Ill.: Biological Computer Laboratory, 1970); Humberto Maturana and Francisco Varela, *Autopoietic Systems: A Characterization of the Living Organization* (Urbana, Ill.: Biological Computer Laboratory, 1975). Regarding the development of the term "autopoiesis," see Humberto Maturana, "The Origin of the Theory of Autopoietic Systems," in *Autopoiesis. Eine Theorie im Brennpunkt der Kritik*, ed. Hans Rudi Fischer (Heidelberg: Carl Auer, 1991), 121–124.

19. Humberto Maturana, Ricardo Uribe, and Francisco Varela, "Autopoiesis: The Organization of Living Systems, Its Characterization and a Model," *Biosystems* 5, no. 4 (1974): 187–196.

20. See Humberto Maturana and Francisco J. Varela, *The Tree of Knowledge: The Biological Roots of Human Understanding*, rev. ed. (Boston: Shambhala Publications, 1998).

21. Heinz von Foerster, "Molecular Ethology, an Immodest Proposal for Semantic Clarification," in *Molecular Mechanisms in Memory and Learning*, ed. Georges Ungar (New York: Plenum Press, 1970), 213–248; UU, 133 ff.

22. Lynn Margulis, *Symbiosis in Cell Evolution: Microbial Communities in the Archean and Proterozoic Eons*, 2nd ed. (New York: W. H. Freeman, 1993).

23. Jacques Monod, *Chance and Necessity: An Essay on the Natural Philosophy of Modern Biology* (New York: Vintage Books, 1972).

24. See Gordon Pask, "The meaning of cybernetics in the behavioural sciences (The cybernetics of behaviour and cognition; extending the meaning of 'goal')," *CoC* 402–416.

25. Wittgenstein, TLP 4.0621: "But it is important that the signs 'p' and '~p' can say the same thing. For it shows that nothing in reality corresponds to the sign '~'."

26. See Gotthard Günther, "Die aristotelische Logik des Seins und die nicht-aristotelische Logik der Reflexion," in *Zeitschrift für philosophische Forschung* 12 (1958): 360–407; Günther, "Ein Vorbericht über die generalisierte Stellen-werttheorie der mehrwertigen Logik," in *Grundlagenstudien* 1, H. 4, (1960), 90–104; Günther, "Cybernetic Ontology and Transjunctional Operations," in *Self-Organizing Systems*, ed. Marshall C. Yovits, George T: Jacobi, and Gordon T. Goldstein (Washington, D.C.: Spartan Books, 1962), 313–392.

27. Alfred North Whitehead and Bertrand Russell, *Principia Mathematica* (Cambridge: Cambridge University Press, 1950).

28. Margulis, *Symbiosis in Cell Evolution*.

29. See Gregory Bateson, *Steps to an Ecology of Mind: Collected Essays in Anthropology, Psychiatry, Evolution, and Epistemology* (London: Jason Aronson, 1972), 9 ff.

30. Heinz von Foerster, "Responsibilities of Competence," *Journal of Cybernetics* 2 (1972): 1–6; UU, 195.

THIRD DAY: MOVEMENT, SPECIES, RECURSION, SELECTIVITY

1. Heinz von Foerster, "Notes on an Epistemology for Living Things," UU.

2. During the week of interviews, the Hale-Bopp comet was visible.

3. See David Hilbert, *Gesammelte Abhandlungen*, vol. 3 (Berlin: Julius Springer, 1936).

4. Mitchell Feigenbaum discovered (or invented) the so-called Feigenbaum-number 4669 as a basic element of recursions in dynamic systems—and as a kind of natural number—through experiments with his calculator.

5. Heinz von Foerster used such an operator in his "Principles of Self-Organization in a Socio-Managerial Context," in *Self-Organization and Manage-*

ment of Social Systems, ed. Hans Ulrich and Gilbert Probst (Berlin: Springer, 1984), 2–24.

6. For an overview, see John Casti, *Reality Rules: Picturing the World in Mathematics* (New York: Wiley, 1992).

7. [The peculiar something that comes out is, in German, "Eigen-Artiges." The term punningly alludes to eigenvalues and to the inventor of hypercycles, Manfred Eigen.—Trans.]

8. See Benoit B. Mandelbrot, *The Fractal Geometry of Nature* (New York: Freeman, 1977).

9. The Swedish mathematician Helge von Koch (1870–1924) was among the first scientists to investigate fractals.

10. See Heinz-Otto Peitgen and Peter H. Richter, *The Beauty of Fractals* (Berlin: Springer, 1986).

11. See Stephen Toulmin, *Human Understanding* (Princeton: Princeton University Press, 1972).

12. Heinz von Foerster, "Thoughts and Notes on Cognition," UU 184.

13. ["*Abbildung*" in German. Foerster is clearly referring to Wittgenstein's concept, which Wittgenstein rendered in English as "picture." While the reference is important, to follow Wittgenstein's word choice would present too many problems in the context of this discussion.—Trans.]

14. See Humberto Maturana and Francisco J. Varela, *The Tree of Knowledge: The Biological Roots of Human Understanding*, rev. ed. (Boston: Shambhala Publications, 1998), 129 ff.

15. "But one thing is the thought, another thing is the deed, and another thing is the idea of the deed. The wheel of causality doth not roll between them," writes Friedrich Nietzsche in *Thus Spoke Zarathustra*.

16. Jean Piaget, *La construction du réel chez l'enfant* (Neuchâtel: Delachaux et Niestlé, 1937).

17. [*Com*, meaning "together" + *prendre* or *prehendere*, "to take, to grasp."—Trans.]

18. Susan Langer, *Philosophy in a New Key: A Study in the Symbolism of Reason, Rite, and Art* (Cambridge, Mass.: Harvard University Press, 1951).

19. See Karl R. Popper, *Objective Knowledge: An Evolutionary Approach* (Oxford: Clarendon Press, 1971), 61 ff.

20. That external stimuli from the world, which is to be perceived, form a necessary but not a sufficient condition for perception—that perception is, instead, the activity of the perceiver—is an idea that we find in Henri Poincaré, "L'éspace et la geometrie," *Revue de Métaphysique et de Morale* 3 (1895): 631–664.

21. [In the original German, Karl and Albert are making a joke about the earlier references to plant movement by punning on the German for "to reproduce, to propagate" *fortpflanzen*, which literally means "plant away/gone." Foerster laughs appreciatively.—Trans.]

22. See Ernst von Glasersfeld, "An Introduction to Radical Constructivism," in *The Invented Reality*, ed. Paul Watzlawick (New York: Norton, 1984), 17–40.

23. Maturana and Varela, *The Tree of Knowledge*.

24. The first volume of Ulyssis Aldrovandi's *Natural History* appeared in 1599.

25. Duarte's is the best restaurant in Pescadero.

26. Carl Linnaeus (1707–1778) was among the greatest scientists of his time. His system of classification revolutionized botany in the eighteenth century.

27. Umberto Eco's novel *The Name of the Rose* ends with the words: "*Stat rosa pristina nomine, nomina nuda tenemus*": "The rose of yore stands only as name; we keep the nude names."

FOURTH DAY: COGNITION, PERCEPTION, MEMORY, SYMBOLS

1. See Olaf Breidbach, *Die Materialisierung des Ichs. Zur Geschichte der Hirnforschung im 19. und 20. Jahrhundert* (Frankfurt: Suhrkamp, 1997), 118 ff.; Antonio R. Damasio, *Descartes' Error: Emotion, Reason and the Human Brain* (New York: Avon Books, 1997), with a description of the case of Phineas P. Gage.

2. Warren S. McCulloch, "Why the Mind Is in the Head," in *Cerebral Mechanisms in Behavior: The Hixon Symposium*, ed. Lloyd A. Jeffress (New York: Wiley, 1951), 42–111.

3. Damasio, *Descartes' Error*; Gerald M. Edelman, *Neural Darwinism: The Theory of Neuronal Group Selection* (New York: Basic Books, 1987); Murray Gell-Mann, *The Quark and the Jaguar: Adventure in the Simple and the Complex* (New York: Freeman, 1994); Douglas R. Hofstadter, *Gödel Escher Bach: An Eternal Golden Braid* (New York: Basic Books, 1979), *Le Ton Beau de Marot: In Praise of the Music of Language* (New York: Basic Books, 1997); John H. Holland, *Hidden Order: How Adaptation Builds Complexity* (Reading, Mass.: Addison Wesley, 1995); Marvin Minsky, *The Society of Mind* (New York: Touchstone, 1988); Daniel C. Dennett, *Content and Consciousness* (London: Routledge, 1986), ix.

4. Humberto Maturana, Gabriele Uribe, and Samy Frenk, "A Biological Theory of Relativistic Color Coding in the Primate Retina," *Archivos de biologia y medicina experimentales*, suppl. 1 (1969).

5. Maturana and Varela, *The Tree of Knowledge*, 132. Almost all such images are based on the visual representations made by the German physician Fritz Kahn (1888–1968).

6. Heinz von Foerster, "Thoughts and Notes on Cognition," in *Cognition: A Multiple View*, ed. Paul L. Garvin (Washington, D.C.: Spartan Books, 1970), 25–48; Humberto R. Maturana, "Neurophysiology of Cognition," ibid., 3–23.

7. See John C. Eccles, *The Neurophysiological Basis of Mind* (Oxford: Clarendon Press, 1953). Ramón y Cajal won the Nobel Prize in 1906.

8. Heinz von Foerster, "On Constructing a Reality," in *Environmental Design Research*, ed. Wolfgang F. E. Preiser (Stroudsburg, Pa.: Dowden, Hutchinson and Ross, 1973), 2:35–46; republished in UU.

9. On the Macy-Conferences, see, besides Heinz von Foerster's own contributions in UU, Steve Joshua Heims, *Constructing a Social Science for Postwar America: The Cybernetics Group 1946–1953* (Cambridge, Mass.: MIT Press, 1993). The conferences have been documented in Heinz von Foerster, ed., *Cybernetics: Transactions of the Sixth Conference* (New York, 1949); Heinz von Foerster, Margaret Mead, and Hans Lukas Teuber, eds., *Cybernetics: Transactions of the Seventh Conference* (New York, 1950); *Cybernetics: Transactions of the Eighth Conference* (New York, 1951); *Cybernetics: Transactions of the Ninth Conference* (New York, 1953); *Cybernetics: Transactions of the Tenth Conference* (New York, 1955). These volumes have been republished under the secondary editorship and name of Claus Pias.

10. On Franz Joseph Gall, see Breidbach, *Die Materialisierung des Ichs*.

11. Annie was our loveable host at the Old Saw Mill Lodge, Pescadero, California.

12. See Heinz Förster, *Das Gedächtnis. Eine quantenphysikalische Untersuchung* (Vienna: Deuticke, 1948).

13. Claude E. Shannon and Warren Weaver, *The Mathematical Theory of Communication* (Urbana: University of Illinois Press, 1949).

14. See Sidney M. Dancoff and Henry Quastler, "The Information Content and Error Rate of Living Things," in *Information Theory in Biology*, ed. Henry Quastler (Urbana: University of Illinois Press, 1953), 263–273.

15. W. Ross Ashby, "Can a Mechanical Chess-player Outplay His Designer," *British Journal for the Philosophy of Science* (1952): 44–57.

16. Heinz von Foerster, "Computation in Neural Nets," *Currents in Modern Biology* 1 (1967): 47–93.

17. See David E. Rumelhart and James L. McClelland, eds., *Parallel Distributed Processing* (Cambridge, Mass.: MIT Press, 1986).

18. Edgar Allan Poe, *Maelzel's Chess Player* (1836).

19. See W. Ross Ashby, "Requisite Variety and its Implications for the Study of Complex Systems," *Cybernetica* 1 (1958): 83–99.

20. See Gregory Bateson, *Steps to an Ecology of Mind: Collected Essays in Anthropology, Psychiatry, Evolution, and Epistemology* (London: Jason Aronson, 1972).

21. Henry Plotkin, *Darwinian Machines and the Nature of Knowledge* (Cambridge, Mass.: Harvard University Press, 1994).

FIFTH DAY: COMMUNICATING, TALKING, THINKING, FALLING

1. Gregory Bateson, "Problems in Cetacean and Other Mammalian Communication," in *Steps to an Ecology of Mind: Collected Essays in Anthropology, Psychiatry, Evolution, and Epistemology* (London: Jason Aronson, 1972), 364–378.

2. Gregory Bateson, "A Theory of Play and Fantasy," in ibid., 177–193.

3. Wittgenstein PI, 327. If a lion could talk, we wouldn't be able to understand it.

4. See Humberto R. Maturana, "The Organization of the Living: A Theory of the Living Organization," *International Journal of Man-Machine-Studies* 7, no 3 (1975): 313–332.

5. Humberto Maturana, Ricardo Uribe, and Francisco Varela, "Autopoiesis: The Organization of Living Systems, Its Characterization and a Model," *Biosystems* 5, no. 4 (1974): 187–196.

6. Erich H. Lenneberg, *Biological Foundations of Language* (New York: Wiley, 1967); "On Explaining Language," *Science* 164 (1969): 635–643.

7. Humberto R. Maturana, "Biology of Language: The Epistemology of Reality," in *Psychology and Biology of Language and Thought: Essays in Honour of Eric H. Lenneberg*, ed. George A. Miller and Elizabeth Lenneberg (New York: Academic Press, 1978), 27–63.

8. Bateson, *Steps to an Ecology of Mind*.

9. Ludwig Wittgenstein was called "uncle" by young Heinz because his mother, Lilith Förster, was a dear friend of Margarethe Stonborough, Ludwig's sister. See Heinz von Foerster, UU; Karl H. Müller, "Wittgensteins Neffe," in *Konstruktivismus und Kognitionswissenschaft. Kulturelle Wurzeln und Ergebnisse. Heinz von Foerster gewidmet*, ed. Albert Müller, Karl H. Müller, and Friedrich Stadler (Vienna: Springer, 1997).

10. See Heinz von Foerster, "Technology: What Will It Mean to Librarians?" *Illinois Libraries* 53 (1971): 785–803; reprinted in OS, 277.

11. See Heinz von Foerster, "Die Magie der Sprache und die Sprache der Magie," in *Abschied von Babylon. Verständigung über Grenzen der Psychiatrie*, ed. Thomas Bock et al. (Bonn: Psychiatrie Verlag, 1995), 24–35.

12. See Joseph Weizenbaum, "ELIZA—A Computer Program for the Study of Natural Language Communication Between Man and Machine," *Communications of the Association for Computing Machinery* 9 (1965): 36–45.

13. See Paul Weston, "To Uncover, To Deduce, To Conclude," *Computer Studies in the Humanities and Verbal Behavior* 3 (1970): 77–89.

14. Heinz von Foerster, "Objects: Tokens for (Eigen-)Behaviors," *ASC Cybernetics Forum* 8, (1976): 91–96. Reprinted in UU.

15. See Francisco J. Varela, "The Ages of Heinz von Foerster," in OS, xiii–xviii.

16. See Noam Chomsky, *Aspects of the Theory of Syntax* (Cambridge, Mass.: MIT Press, 1965).

17. See Ernst von Glasersfeld, *Radical Constructivism: A Way of Knowing and Learning* (London: Routledge, 1995).

18. Heinz von Foerster, "On Self-Organizing Systems and Their Environments."

19. See Steven Pinker, *The Language Instinct* (New York: Morrow, 1994).

20. Ibid., 18.

21. Foerster often handed out such lists during his lectures.

22. Ernst Mach, *The Analysis of Sensations, and the Relation of the Physical to the Psychical* (New York: Dover Publications, 1959).

23. See Douglas Hofstadter, *Gödel Escher Bach: An Eternal Golden Braid* (New York: Basic Books, 1976).

24. Heinz von Foerster presented us with an abridged and also sharpened version of this story. See Bertrand Russell, *The Autobiography of Bertrand Russell, 1872–1914* (London: Routledge, 1967), 147 ff.

25. See Kurt Gödel, "Über formal unentscheidbare Sätze der Principia Mathematica und verwandter Systeme I," *Monatshefte für Mathematik und Physik* 38 (1931): 173–198.

26. Spencer Brown, *Laws of Form*. Heinz von Foerster wrote an early review of Spencer Brown's book in the *Whole Earth Catalog* (Sausalito, Calif.: Portola Institute, 1970), 14.

27. Michael S. Gazzaniga, *The Social Brain: Discovering the Networks of the Mind* (New York: Basic Books, 1985). The experiment is presented on page 71.

28. "I offer you the phrase *the pattern which connects* as a synonym, another possible title for this book. *The pattern which connects.* . . . What pattern connects the crab to the lobster and the orchid to the primrose and all the four of them to me? And me to you? And all the six of us to the amoeba in one direction and to the back-ward schizophrenic in another?" Gregory Bateson, *Mind and Nature: A Necessary Unity* (New York: Dutton, 1979), 8.

29. On the notion of deutero-learning, see Bateson, *Steps to an Ecology of Mind.*

30. Massimo Piattelli-Palmarini, ed., *On Language and Learning: The Debate Between Jean Piaget and Noam Chomsky* (Cambridge, Mass.: Harvard University Press, 1980).

31. See Keith Lehrer and Carl Wagner, *Rational Consensus in Science and Society: A Philosophical and Mathematical Study* (Dordrecht: Reidel, 1991).

32. Thomas S. Kuhn, *The Structure of Scientific Revolutions*, 2nd ed. (Chicago: University of Chicago Press, 1970).

33. Ludwik Fleck, *Genesis and Development of a Scientific Fact* (Chicago: University of Chicago Press, 1979). (First publication in German 1935.)

34. Wittgenstein, PI § 83.

35. Heinz von Foerster, "Computation in Neural Nets," *Currents in Modern Biology* 1 (1967): 47–93. Reprinted in UU.

36. David E. Rumelhart and James L. McClelland, eds., *Parallel Distributed Processing*, 2 vols. (Cambridge, Mass.: MIT Press, 1986).

37. René Thom, *Structural Stability and Morphogenesis: An Outline of a General Theory of Models* (Reading, Mass.: Addison-Wesley, 1989).

1. See Ferdinand Scheminzky, "Kann Leben künstlich erzeugt werden?" in *Alte Probleme—Neue Lösungen in den exakten Wissenschaften* (Leipzig and Vienna, 1934), 67–92.

2. See Friedrich Stadler, *The Vienna Circle: Studies in the Origin, Development, and Influence of Logical Empiricism* (New York: Springer, 2001).

3. Presentations on the significance of Euclid's (fifth) "parallel postulate" and on establishing a non-Euclidian geometry are found in Rudolf Carnap, *Philosophical Foundation of Physics: An Introduction to the Philosophy of Science* (New York: Basic Books, 1966), 125 ff., or in Hans Reichenbach, *The Philosophy of Space and Time* (New York: Dover Publications, 1958).

4. The original claim is "a map is not the territory." See Alfred Korzybski, "A Non-Aristotelian System and Its Necessity Rigour in Mathematics and Physics," in *Science and Sanity: An Introduction to Non-Aristotelian Systems and General Semantics*, 5th ed. (Brooklyn, N.Y.: Institute of General Semantics, 1994), 747–761, here 750.

5. See the older research by Heinz von Foerster, G. Brecher, and E. Cronkite, "Produktion, Ausreifung und Lebensdauer der Leukozyten," in *Physiologie und Physiopathologie der weissen Blutzellen*, ed. H. Braunsteiner (Stuttgart: Georg Thieme, 1959), 188–214; "Some Remarks on Changing Populations," in *The Kinetics of Cellular Proliferation*, ed. F. Stohlman Jr. (New York, 1959), 382–407; Heinz von Foerster, Patricia M. Mora, and Lawrence W. Amiot, "Doomsday," *Science* 133 (1961): 936–946; "Population Density and Growth," *Science* 133 (1961): 1931–1937. On this very interesting demographic work, see Stuart A. Umpleby, "The Scientific Revolution in Demography," in *Population and Environment: A Journal of Interdisciplinary Studies* 11 (1990): 159–174.

6. Otto Neurath, *Empiricism and Sociology*, ed. Marie Neurath and Robert S. Cohen (Dordrecht: Reidel, 1973).

7. On the terms "Modus I" and "Modus II," see especially Michael Gibbons, Camille Limoges, Helga Nowotny, et al., *The New Production of Knowledge: The Dynamics of Science and Research in Contemporary Societies* (London: Sage, 1994).

8. Humberto R. Maturana and Francisco Varela, *The Tree of Knowledge: The Biological Roots of Human Understanding* (Boston: Shambhala, 1992).

9. Heinz von Foerster, "Cybernetics of Cybernetics," in *Communication and Control in Society*, ed. Klaus Krippendorf (New York: Gordon and Breach, 1979), 5.

10. "Notes on an Epistemology for Living Things."

11. "When Banzan was walking through a market, he overheard a conversation between a butcher and his costumer. 'Give me the best piece of meat you have,' said the customer. 'Everything in my shop is the best,' replied the butcher. 'You cannot find here any piece of meat that is not the best.' At these words

Banzan became enlightened." Paul Reps, *Zen Flesh, Zen Bones* (New York: Doubleday, 1972).

12. [While literally meaning "pattern solution," *Musterlösung* would normally be translated as "model solution."—Trans.]

13. The Naval Research Office sponsored and supported—among others—the following meetings and conferences: Marshall C. Yovits and Scott Cameron, eds., *Self-Organizing Systems* (New York: Pergamon, 1960); Marshall C. Yovits, George T. Jacobi, and Gordon D. Goldstein, eds., *Self-Organizing Systems* (Washington, D.C.: Spartan Books, 1962); Heinz von Foerster and G. W. Zopf Jr., eds., *Principles of Self-Organization: The Illinois Symposium on Theory and Technology of Self-Organizing Systems* (London: Pergamon, 1962).

14. In the 1960s, Roger W. Sperry was Hixon Professor of Psychobiology at the California Institute of Technology (CalTech). See Roger W. Sperry, "The Growth of Nerve Circuits," *Scientific American* 201 (1959): 68–75; "The Corpus Callosum and Interhemispheric Transfer in the Monkey," *Anatomical Record* 131 (1958): 297; "Cerebral Organization and Behavior," *Science* 133 (1961): 1449–1457.

15. Heinz von Foerster, "Ethique et cybernétique de second ordre," in *Systèmes, ethique, perspectives en thérapie familiale*, ed. Yveline Ray and Bernard Prieur (Paris: Edition ESF, 1991), 41–55; "Ethics and Second Order Cybernetics," UU, 287–304.

16. See Niklas Luhmann, *Die Gesellschaft der Gesellschaft*, 2 vols. (Frankfurt: Suhrkamp, 1997).

17. Regarding Stafford Beer, see his books republished as the Stafford Beer Classic Library, among them *Decision and Control: The Meaning of Operational Research and Management Cybernetics* (Chichester: Wiley, 1994); *The Brain of the Firm* (Chichester: Wiley, 1994); and *The Heart of the Enterprise* (Chichester: Wiley, 1994).

18. Fernando Flores later returned to the United States, where he became head of several computer and software companies. Still most interesting to read is a book he wrote together with Terry Winograd, *Understanding Computers and Cognition: A New Foundation for Design* (Reading, Mass.: Addison-Wesley, 1986).

19. A possible motto that connects first-level operators—and by chance there are seven of them—might read: "Grimmian Jacobins in manifold role-types [produce, fabricate, create, generate . . .] pattern solutions recursively before a large audience." In Heinz von Foerster's shortened version, they invite us into "a permanent dance with the world." A possible motto connecting the second level operators—[From] [Here and now] [I] [Orders]—might be created in the following way: "Here I form orders," in which all of the operators are localized on the second level and apply to themselves: [The Here and Now of the Here and Now . . .] [Form of Form . . .] [I of I . . .] [Order of Order . . .]. This previous construction becomes most interesting because of how very well it fits with

Douglas R. Hofstadter's characterization of "creative programs": "Full-scale creativity consists in having a keen sense for what is interesting, following it recursively, applying it at the meta-level, and modifying it accordingly." Douglas R. Hofstadter, *Fluid Concepts and Creative Analogies: Computer Models of the Fundamental Mechanisms of Thought* (New York: Basic Books, 1995), 313. These second-level operators are able—and this is an important clue toward the construction of the Foerster thought machine—to connect effortlessly to that sought-after metalevel that gets the first-level operators—invert, unite, move, etc.—going recursively in a certain direction.

20. Wittgenstein, *Culture and Value*, 50.

SEVENTH DAY: REST, REST, REST, REST

1. The Heinz von Foerster Clock has been constructed from propositions of "Notes on an Epistemology for Living Things."

INDEX

Lightning Source UK Ltd.
Milton Keynes UK
UKHW011829080622
404134UK00001B/113